Lecture Notes in Mathematics

A collection of informal reports and seminars
Edited by A. Dold, Heidelberg and B. Eckmann, Zürich

174

Birger Iversen

Linear Determinants with Applications to the Picard Scheme of a Family of Algebraic Curves

Springer-Verlag
Berlin · Heidelberg · New York

This series aims to report new developments in mathematical research and teaching – quickly, informally and at a high level. The type of material considered for publication includes:

1. Preliminary drafts of original papers and monographs
2. Lectures on a new field, or presenting a new angle on a classical field
3. Seminar work-outs
4. Reports of meetings

Texts which are out of print but still in demand may also be considered if they fall within these categories.

The timeliness of a manuscript is more important than its form, which may be unfinished or tentative. Thus, in some instances, proofs may be merely outlined and results presented which have been or will later be published elsewhere.

Publication of *Lecture Notes* is intended as a service to the international mathematical community, in that a commercial publisher, Springer-Verlag, can offer a wider distribution to documents which would otherwise have a restricted readership. Once published and copyrighted, they can be documented in the scientific literature.

Manuscripts

Manuscripts are reproduced by a photographic process; they must therefore be typed with extreme care. Symbols not on the typewriter should be inserted by hand in indelible black ink. Corrections to the typescript should be made by sticking the amended text over the old one, or by obliterating errors with white correcting fluid. Should the text, or any part of it, have to be retyped, the author will be reimbursed upon publication of the volume. Authors receive 75 free copies.

The typescript is reduced slightly in size during reproduction; best results will not be obtained unless the text on any one page is kept within the overall limit of 18 x 26.5 cm (7 x 10 ½ inches). The publishers will be pleased to supply on request special stationery with the typing area outlined.

Manuscripts in English, German or French should be sent to Prof. Dr. A. Dold, Mathematisches Institut der Universität Heidelberg, Tiergartenstraße or Prof. Dr. B. Eckmann, Eidgenössische Technische Hochschule, Zürich.

Die „*Lecture Notes*" sollen rasch und informell, aber auf hohem Niveau, über neue Entwicklungen der mathematischen Forschung und Lehre berichten. Zur Veröffentlichung kommen:

1. Vorläufige Fassungen von Originalarbeiten und Monographien.
2. Spezielle Vorlesungen über ein neues Gebiet oder ein klassisches Gebiet in neuer Betrachtungsweise.
3. Seminarausarbeitungen.
4. Vorträge von Tagungen.

Ferner kommen auch ältere vergriffene spezielle Vorlesungen, Seminare und Berichte in Frage, wenn nach ihnen eine anhaltende Nachfrage besteht.

Die Beiträge dürfen im Interesse einer größeren Aktualität durchaus den Charakter des Unfertigen und Vorläufigen haben. Sie brauchen Beweise unter Umständen nur zu skizzieren und dürfen auch Ergebnisse enthalten, die in ähnlicher Form schon erschienen sind oder später erscheinen sollen.

Die Herausgabe der „*Lecture Notes*" Serie durch den Springer-Verlag stellt eine Dienstleistung an die mathematischen Institute dar, indem der Springer-Verlag für ausreichende Lagerhaltung sorgt und einen großen internationalen Kreis von Interessenten erfassen kann. Durch Anzeigen in Fachzeitschriften, Aufnahme in Kataloge und durch Anmeldung zum Copyright sowie durch die Versendung von Besprechungsexemplaren wird eine lückenlose Dokumentation in den wissenschaftlichen Bibliotheken ermöglicht.

Lecture Notes in Mathematics

A collection of informal reports and seminars
Edited by A. Dold, Heidelberg and B. Eckmann, Zürich

174

Birger Iversen
MIT, Cambridge, MA/USA

Linear Determinants with Applications to the Picard Scheme of a Family of Algebraic Curves

Springer-Verlag
Berlin · Heidelberg · New York 1970

ISBN 3-540-05301-8 Springer-Verlag Berlin · Heidelberg · New York
ISBN 0-387-05301-8 Springer-Verlag New York · Heidelberg · Berlin

Offsetdruck: Julius Beltz, Weinheim/Bergstr.

CONTENTS

Introduction.. IV

I. The linear determinant.............................. 1

II. Representation of n-fold sections by symmetric products.. 20

III. Invertibles sheaves and rational maps into $C^{(g)}$.. 32

IV. Construction of the Picard scheme of a family of curves.. 52

Bibliography.. 69

INTRODUCTION

This paper grew out of a study of A. Weil's construction of the Jacobian variety of an algebraic curve.

The first delicate problem in Weil's construction is a rationality question connected with the symmetric product of a curve, which led me to the theory of linear determinants as exposed in Chapter I. This theory studies a 1-dimensional integral representation, called ld , of $M_{n,Z}^{(n)}$ which denotes the n-fold symmetric product of the $n \times n$ matrices i.e. the algebra of tensors in $M_{n,Z} \otimes \ldots \otimes M_{n,Z}$ (n factors) invariant under the standard action of the symmetric group on n letters. This representation aside from its geometric aspects provides a linearization of Galois theory different from the one usually advocated. A relation of ld to Azumaya algebras is just touched upon. I have kept this chapter entirely in the notion of commutative algebra since I hope it has an independent interest.

In chapter II is studied the geometric aspect of ld which is first of all the following: Let $f : X \to Y$ be a finite morphism of schemes whose fibers have constant rank n and let $X_Y^{(n)}$ denote the n-fold symmetric product of X over Y . By means of ld is constructed a section of $X_Y^{(n)}$ over Y which underlies the geometric map which to a geometric point y of Y associates the unordered

*This work was partially supported by Air Force Contract No. F44620-67-C-0029.

n-tuple $\{x_1,\ldots,x_n\}$ where $x_1,\ldots,x_n$ denotes the geometric points of X lying over y each repeated as many times as the ramification index prescribes. This is applied to prove that the n-fold symmetric product of a flat family of non singular curves represents the functor "n-fold sections". This theorem solves the rationality question mentioned above. In case of a projective family of curves "n-fold sections" becomes relative Cartier-divisors and the theorem ties up with the theory of Chow points and Grothendiecks theory of Hilbert Schemes.

The goal of Chapter III and IV is to generalize Weil's construction of the Jacobian variety of a single curve to construct Picard schemes, in the sense of Grothendieck, for a flat and proper family of geometrically reduced and irreducible curves.*) The result we get is that the Picard scheme of such a family exists after a faithfully flat, finite type extension of the parameter scheme. Recent results of M Raynaud [Ra]**) p. 178 allow to descent the obtained Picard schemes in the case where either the parameter scheme is 1-dimensional or the case where all fibers are non singular. Hopefully, future results along the lines

*) The additional technical assumptions are: the base scheme is noetherian and the family satisfies condition II.1.1.

**) Square brackets contain references to the Bibliography at the end of the paper.

of Raynaud will allow to descent the result in general.

In case of a projective family of curves the result is a special case of Grothendieck's general existence theorem for the picard schemes of a projective family.

The main tools in the construction given here are M. Artin's generalization of Weil's theorem on the construction of a group from a rational group law and of course Grothendieck's theorems on cohomology of coherent sheaves, especially those of base change type. It should also be mentioned that the possibility of allowing singular fibers in this construction was opened up by theorems of Rosenlicht $[R_2]$. Finally a new (I believe) version of the seesaw principle is instrumental, a proof of this version in its full generality is included in Chapter IV.

This material has been presented at a seminar at M.I.T.; as a result I have broadened the paper with a few proofs of older results due to Weil and others. I would like to take the opportunity to thank Connie Clayton for her careful and fast typing of the manuscript.

Cambridge, Massachusetts

April 1970

B.I.

CHAPTER I

THE LINEAR DETERMINANT

I.1 Symmetric products and flatness

I.2 Construction of the linear determinant

I.3 Hamilton-Cayley factorization of the linear determinant

I.4 Representations induced by the linear determinant

Appendix. Remarks on the structure of $M_{n,\mathbb{Z}}^{(n)}$.

I.1. Symmetric products and flatness.

Let A be a commutative ring and M an A-module. The symmetric group on n letters, S_n acts on $\otimes_A^n M$ by $\sigma(\otimes_{i=1}^n m_i) = \otimes_{i=1}^n m_{\sigma^{-1}(i)}$. The A-module of invariant tensors will be denoted $M_A^{(n)}$.

If $v \in \otimes_A^n M$ then $\square v$ denotes the sum of the elements in the orbit of v under S_n . This notation will be used whenever we have a linear action of S_n on an abelian group.

Proposition 1.1. If M is a flat A-module then so is $M_A^{(n)}$.

Proposition 1.2. If M is a flat A-module and $A \to B$ a ring homomorphism then the canonical map

$$M_A^{(n)} \otimes_A B \to (M \otimes_A B)_B^{(n)}$$

is an isomorphism.

Proof: According to a theorem of Daniel Lazard, [L], Th. 1.2, p. 84, a flat A-module is direct limit over a filtered set of finitely generated, free A-modules.

It is easy to see that such limits are preserved by the functor $M \mapsto M_A^{(n)}$. It remains to prove 1.1 and 1.2 in case M is a free A-module. But if $(m_i)_{i\in I}$ is a basis for M and we for $f\in I^n$ put

$$m_f = \otimes_{i=1}^{n} m_{f(i)}$$

then $(\Box m_f)$ as f runs through I^n/S_n form a basis for $M_A^{(n)}$. 1.1 and 1.2 now follows immediately.

Q.E.D.

Essentially the same proof will establish

<u>Lemma 1.3</u>. Let M be a flat A-module, N any A-module. Then the canonical map

$$M_A^{(n)} \otimes_A N \to (\otimes_A^n M) \otimes_A N$$

is injective and identifies $M_A^{(n)} \otimes N$ with the invariants of $(\otimes_A^n M) \otimes N$ under the induced action of S_n.

In particular if C is a flat A-module we have a canonical map

1.4. $$C_A^{(n+m)} \to C_A^{(n)} \otimes C_A^{(m)}$$

<u>PROPOSITION 1.5</u>. Let C be an associative A-algebra which is flat as an A-module. Then the canonical map

$$C_A^{(n)} \otimes_A C \to C_A^{(n-1)} \otimes_A C$$

(i.e. the map induced by

$$x_o \otimes \ldots \otimes x_{n-1} \otimes x_n \to x_o \otimes \ldots \otimes x_{n-1} x_n)$$

is surjective.

<u>Proof</u>: Consider the above map as a map of right C-modules. By Lazard's theorem [L] Th. 1.2, p. 86, $C_A^{(n-1)} \otimes_A C$ is a direct limit (over a filtered set) of C-modules of the form $L_A^{(n-1)} \otimes_A C$ where L is a free A-module.

It follows that the C-module $C^{(n-1)} \otimes_A C$ is generated by tensors of the form

$$((a_i)_{i\in I}, f) = \sum_{g, g \equiv f} a_{g(1)} \otimes \ldots \otimes a_{g(n-1)} \otimes 1$$

where $(a_i)_{i\in I}$ is a family of elements of C, $f \in I^{n-1}$ and the sum being over all $g \in I^{n-1}$ which are in the orbit of f under S_{n-1}. Note that $((a_i)_{i\in I}, f)$ is in general different from $(\sqcup a_{f(1)} \otimes \ldots \otimes a_{f(n)}) \otimes 1$. Let I_e denote the set obtained from I by adding an extra element e and let $(a_i)_{i\in I_e}$ denote the prolongation of $(a_i)_{i\in I}$ to I_e with $a_e = 1$. With this notation we want to prove by induction on $m \geq 1$ that the image of $C^{(n)} \otimes C \to C^{(n-1)} \otimes C$ contains all tensors of the form $((a_i)_{i\in I_e}, f)$ where $f \in I_e^{n-1}$ and Card $f^{-1}(e) \geq n-m$. This is clear for $m = 1$. $m \to m+1$:

Let f' denote the prolongation of f to $[1,n]$ ith $f'(n) = e$. The image of $((a_i)_{i\in I_e}, f')\otimes 1 \in C^{(n)}\otimes C$ s

$$\sum_{\substack{g, g\equiv f\\ g(n)=e}} a_{g(1)}\otimes\ldots\otimes a_{g(n-1)}\otimes 1 + \sum_{\substack{g, g\equiv f\\ g(n)\neq e}} a_{g(1)}\otimes\ldots\otimes a_{g(n-1)}\otimes a_{g(n)}$$

e first sum is $((a_i)_{i\in I_e}, f)$, the second belongs to e image by the induction hypothesis.

Q.E.D.

emark 1.6. Let C be as in 1.5. The kernel of $C^{(n)}\otimes C \to C^{(n-1)}\otimes C$ contains some prominant tensors:

Let $a\in C$ and let $s_i(a)\in C^{(n)}$ be defined the following identity in $C^{(n)}\otimes_A A[T]$

$$\prod_{i=1}^{n} (T - \underbrace{1\otimes\ldots\otimes a}_{i}\otimes\ldots\otimes 1)$$

$$= T^n + \sum_{i=1}^{n} (-1)^i s_i(a) T^{n-i}$$

n $C^{(n)}\otimes C$ we have the element

$$hc(a) = a^n + \sum_{i=1}^{n} (-1)^i s_i(a) a^{n-i}$$

ere we have identified a with $1\otimes\ldots\otimes 1\otimes a$. It is asy to see that $hc(a)$ is in the kernel of $C^{(n)}\otimes C \to C^{(n-1)}\otimes C$.

I.2. Construction of the linear determinant.

In this section we construct a representation

$$\mathrm{ld} : M_{n,\mathbb{Z}}^{(n)} \to \mathbb{Z}$$

i.e. a ring homomorphism with the property that for $a \in M_{n,\mathbb{Z}}$

2.1. $$\mathrm{ld}(\underbrace{a\otimes \ldots \otimes a}_{n}) = \det a$$

Construction*): Let $e_{r,s}$ denote the $n \times n$ matrix given by $(e_{r,s})_{i,j} = \delta_{r,i}\delta_{s,j}$ (Kronecker delta). Recall that $e_{r,s}e_{t,u} = \delta_{s,t}e_{r,u}$. Let T_n denote the set of maps from $[1,n]$ to $[1,n]$ and for $(f,g) \in T_n^2$ put

$$E_{f,g} = \bigotimes_{i=1}^{n} e_{f(i),g(i)}$$

The $E_{f,g}$'s form a basis for $\otimes_{\mathbb{Z}}^{n} M_{n,\mathbb{Z}}$ as (f,g) runs through T_n^2 and the $\square E_{f,g}$'s form a basis for $M_{n,\mathbb{Z}}^{(n)}$ as (f,g) runs through $T_n^{\,2}/S_n$, the action of S_n on T_n^2 being $\sigma(f,g) = (f\sigma^{-1}, g\sigma^{-1})$. Let ld denote the $\mathbb{Z}$-linear map $M_{n,\mathbb{Z}}^{(n)} \to \mathbb{Z}$ given by

2.2. $$\mathrm{ld}(\square E_{f,g}) = \epsilon_f \epsilon_g$$

---

*) For a different approach see Appendix to Chap. I.

ϵ_h, for $h \in T_n$, is zero if $h \notin S_n$ and equals the signature of h if $h \in S_n$.

Let us verify

$$\mathrm{ld}(\Box E_{f,g} \cdot \Box E_{h,k}) = \mathrm{ld}(\Box E_{f,g})\,\mathrm{ld}(\Box E_{h,k})$$

At our disposal we have

$$E_{r,s}E_{t,u} = \delta_{s,t}E_{r,u}$$

The case where $f \notin S_n$ or $k \notin S_n$ is easily handled. So we assume $f, k \in S_n$. Without loss of generality we may even assume $f = k = e$, the identity permutation. By the just quoted formula one verifies

$$\Box E_{e,g} \Box E_{h,e} = \sum_{\sigma \in S_n} \delta_{g\sigma,h} \Box E_{\sigma,e}$$

So the formula we want to prove is

$$\epsilon_g \epsilon_h = \sum_{\sigma \in S_n} \delta_{g\sigma,h} \epsilon_\sigma$$

If $g\sigma \neq h$ for all $\sigma \in S_n$ then $g \notin S_n$ or $h \notin S_n$ and the formula is obviously true. In case $g\sigma_0 = h$ the formula reduces to

$$(\epsilon_h)^2 = \sum_{\sigma \in S_n} \delta_{h\sigma,h} \epsilon_\sigma$$

the proof of which we leave to the reader.

Let us now verify 2.1: Put $a = (a_{r,s})$ and for $(f,g) \in T_n^2$ put $a_{f,g} = \pi_{i=1}^{n} a_{f(i),g(i)}$, clearly

$a_{f\sigma,g\sigma} = a_{f,g}$ for $\sigma \in S_n$.

$$\underbrace{a\otimes \dots \otimes a}_{n} = \sum_{(f,g)\in T_n^2} a_{f,g} E_{f,g}$$

$$= \sum_{(f,g)\in T_n^2/S_n} a_{f,g} \square E_{f,g}$$

and consequently

$$\mathrm{ld}(\underbrace{a\otimes \dots \otimes a}_{n}) = \sum_{\sigma\in S_n} \epsilon_\sigma a_{e,\sigma} = \det a$$

Q.E.D.

Let A be a commutative ring. Since the canonical map $M_{n,\mathbb{Z}}^{(n)} \otimes_{\mathbb{Z}} A \to M_{n,A}^{(n)}$ is an isomorphism we get the induced representation

2.2. $$\mathrm{ld}_A : M_{n,A}^{(n)} \to A$$

ld_A will usually be written just ld.

It is clear from the proof of 2.1 that

2.3. $$\mathrm{ld}_A(\underbrace{a\otimes \dots \otimes a}_{n}) = \det a$$

whenever $a \in M_{n,A}$. In particular

$$ld_{A[T]}(\underbrace{(T-a)\otimes \ldots \otimes (T-a)}_{n}) = \det(T-a)$$

or otherwise expressed:

The characteristic polynomial of $a \in M_{n,A}$ is

2.4. $$\chi_a(T) = T^n + \sum_{i=1}^{n} (-1)^i \, ld(s_i(a))T^{n-i}$$

where $s_i(a) \in M_{n,A}^{(n)}$ is defined in 1.6. especially

2.5. $$ld(a\otimes 1 \ldots \otimes 1 + 1\otimes a \ldots \otimes 1 + \ldots + 1\otimes 1 \ldots \otimes a) = Tr(a) .$$

2.6. Upper triangular matrices

Let $T_{n,A}$ denote the A-algebra of upper triangular $n \times n$ matrices with coefficients in A and let $\Delta_i : T_{n,A} \to A$ denote the representation which to the matrix $(a_{r,s})_{r \leq s}$ associates a_{ii}.

$$x_1 \otimes x_2 \ldots \otimes x_n \to \prod_{i=1}^{n} \Delta_i(x_i)$$

induces a representation $T_{n,A}^{(n)} \to A$ which is seen to coincide with the restriction of $ld : M_{n,A}^{(n)} \to A$ to $T_{n,A}^{(n)}$.

I.3. Hamilton-Cayley factorization of the linear determinant.

Proposition 3.1. Let A be a commutative ring

$$\mathrm{ld}_A \otimes 1 : M_{n,A}^{(n)} \otimes_A M_{n,A} \to M_{n,A}$$

can be factored through the canonical equimorphism (I.1.4)

$$M_{n,A}^{(n)} \otimes M_{n,A} \to M_{n,A}^{(n-1)} \otimes M_{n,A}$$

Proof. Consider $M_{n,A}^{(n-1)} \otimes M_{n,A}$ and $M_{n,A}$ as right $M_{n,A}$-modules in the obvious way. We are first going to define a $M_{n,A}$-linear map

$$\mathrm{ld}' : M_{n,A}^{(n-1)} \otimes_A M_{n,A} \to M_{n,A}$$

Notation: $T_{n-1,n}$ denotes the set of maps from $[1,n-1]$ to $[1,n]$. If $f \in T_{n-1,n}$ is injective then $\bar{f}$ denotes the element of S_n whose restriction to $[1,n-1]$ is f. Furthermore, we will use the notation introduced in I.2.

For $(f,g) \in T_{n,n-1}^2$ we define

$$\mathrm{ld}'(\square E_{f,g}) = \begin{cases} \epsilon_{\bar{f}}\,\epsilon_{\bar{g}}\, e_{\bar{g}(n),\bar{f}(n)} & \text{if } f \text{ and } g \text{ both are injective} \\ 0 & \text{otherwise} \end{cases}$$

Let φ denote the canonical map I.5

$M_{n,A}^{(n)} \otimes_A M_{n,A} \to M_{n,A}^{(n-1)} \otimes M_{n,A}$. We want to prove that

$1d \otimes 1 = 1d' \varphi$ i.e. that if $(f,g) \in T_n^2$ then

3.2. $$1d'(\varphi(\square E_{f,g} \otimes 1)) = \epsilon_f \epsilon_g$$

This is easy to see in case f is not bijective. So we may assume $f \in S_n$ or even that $f = e$, the identity.

It is not difficult to verify the formula

3.3. $$\varphi(\square E_{e,g} \otimes 1) = \sum_{i=1}^{n} \square E_{\tau_i, g\tau_i} \otimes 1$$

where $\tau_i \in T_{n-1,n}$ denotes the restriction of the transposition (i,n) to $[1,n-1]$.

It is now an easy matter to verify 3.2 by means of 3.3 in case where $g \in S_n$ and in case where g takes less than n-1 distinct values.

Suppose g takes all values but q and that $g(r) = q(s)$, $1 \leq r < s \leq n$, then we get from 3.3

$$1d'(\varphi(\square E_{e,g} \otimes 1)) = -\epsilon_{\overline{g\tau_r}} e_{q,g(r)} - \epsilon_{\overline{g\tau_s}} e_{q,g(s)}$$

but $\overline{g\tau_r} = \overline{g\tau_s} \circ (r,s)$ and $g(r) = g(s)$.

Q.E.D.

The factorization of $1d \otimes 1$ given in 3.1 got its name the following way:

Let $a \in M_{n,A}$, the tensor

$$\overbrace{1\otimes\ldots\otimes 1}^{n}\otimes a^n + \sum_{i=1}^{n} (-1)^i s_i(a)\otimes a^{n-i}$$

is in the kernel of

$$M_{n,A}^{(n)}\otimes M_{n,A} \to M_{n,A}^{(n-1)}\otimes M_{n,A}$$

according to 1.6. Consequently

$$a^n + \sum_{i=1}^{n} (-1)^i \mathrm{ld}(s_i(a))a^{n-i} = 0$$

which is the theorem of Hamilton-Cayley by 2.4.

.4. Representations induced by ld .

Let A be a commutative ring and M a free A-module of finite rank n . An A-basis for M will identify $\mathrm{End}_A(M)$ with $M_{n,A}$ and consequently define a representation

.1. $$\mathrm{ld} : \mathrm{End}_A(M)_A^{(n)} \to A$$

which is easily seen to be independent of the chosen basis. This construction extends easily to the case where M is locally free of rank n .

According to Auslander and Goldmann's generalization of the Skolem-Noether theorem, any automorphism of $M_{n,A}$ is locally interior [A-G], Th. 3.6. So if B is an A-algebra which is isomorphic to $M_{n,A}$ then we have a canonical representation

.2. $$\mathrm{ld} : B_A^{(n)} \to A$$

This is more generally true if B is an Azumaya-algebra of rank n^2 , i.e. becomes isomomphic to $M_{n,A}$ after a faithfully flat base extension, see [A-G] and [G] for details, as it follows immediately from the theory of faithfully flat descent [S.G.A. 60-61] VIII, Cor. 1.2 de Th. 1.1.

<u>Commutative case.</u>

Let A be a commutative ring, B a commutative A-algebra which is locally free of rank n as an A-module. The canonical representation $B \to \mathrm{End}_A(B)$ induces

$$B_A^{(n)} \to \mathrm{End}_A(B)_A^{(n)}$$

which composed by ld from 4.1 gives an A-algebra homomorphism

4.3. $$\theta_{B/A} : B_A^{(n)} \to A$$

which will be called the canonical section of the A-algebra $B_A^{(n)}$.

Let me now list the three basic properites of $\theta_{B/A}$:

4.4. $\theta_{B/A} \otimes 1 : B_A^{(n)} \otimes_A B \to B$ factors through the isomorphism

$$B_A^{(n)} \otimes_A B \to B_A^{(n-1)} \otimes_A B$$

This is a consequence of 3.1.

4.5. Let $\begin{array}{ccc} A & \to & B \\ f\downarrow & & g\downarrow \\ C & \to & D \end{array}$ be a commutative diagram in the category of commutative rings, such that B (resp. D) is locally free of rank n as A-module (resp. C-module). Then the diagram

$$\begin{array}{ccc} B_A^{(n)} & \xrightarrow{\theta_{B/A}} & A \\ {\scriptstyle g^{(n)}}\downarrow & & \downarrow{\scriptstyle f} \\ D_C^{(n)} & \xrightarrow{\theta_{D/C}} & D \end{array}$$

is commutative.

4.6. Let B be the direct product of n-copies of A and let $p_i : B \to A$ denote the ith projection. $\theta_{B/A}$ is induced by

$$\otimes_A^n B \ni x_1 \otimes \ldots \otimes x_n \to \pi_{i=1}^n p_i(x_i)$$

as it follows from 2.6.

Case of a field

Proposition 4.7. Let k be an algebraically closed field and A a finite commutative k-algebra of rank n. Let $\sigma_1, \ldots, \sigma_n$ denote the k-algebra homomorphisms of A into k and let n_i denote the k-rank of the local ring of A at σ_i. Then the following diagram is commutative

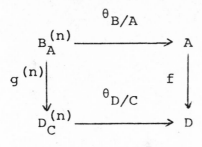

$$\begin{array}{ccc} A_k^{(n)} & \xrightarrow{\theta_{A/k}} & k \\ \big\downarrow & \nearrow_{\sigma} & \\ \otimes_k^n A & & \end{array}$$

where $\sigma = (\underbrace{\sigma_1,\ldots,\sigma_1}_{n_1}, \underbrace{\sigma_2,\ldots,\sigma_2}_{n_2},\ldots\ldots, \underbrace{\sigma_r,\ldots,\sigma_r}_{n_r})$

<u>Proof</u>: $\theta_{A/k}$ can be extended to $\otimes_k^n A$ be Cohen-Seidenberg's theorem. Let $\tau = (\tau_1,\ldots,\tau_n)$ be an extension. Let A_i denote the local ring of A at σ_i and $p_i : A \to A_i$ the canonical projection. Let $e_i \in A$ be such that $p_i(e_i) = \delta_{i,j}$.

The characteristic polynomial for e_i is

$$\chi_{e_i}(X) = (X-1)^{n_i} X^{n-n_i}$$

An alternative way of computing χ_{e_i} is

$$\chi_{e_i}(X) = \theta_{A[X]/k[X]}((X-e_i)\otimes\ldots\otimes(X-e_i)) = \prod_{j=1}^{n}(X-\tau_j(e_i))$$

Consequently, $\tau_j = \sigma_i$ for n_i values of j.

Q.E.D.

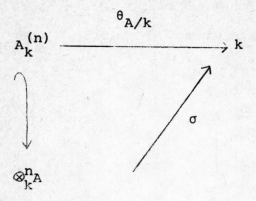

orollary 4.8. (of the proof). If $a \in A$ then the oots of the characteristic polynomial of a are

$\sigma_i(a)$ with multiplicity n_i, $i = 1,\dots,r$.

orollary 4.9. Let k be an arbitrary field, A a inite commutative k-algebra of rank k, $\sigma_1,\dots,\sigma_n$ the he k-morphisms from A to $\bar{k}$ (repeated with the multilicities defined in Prop. 4.7), $\sigma = (\sigma_1,\dots,\sigma_n) : \otimes_k^n A \to \bar{k}$. f $t \in \otimes_k^n A$ is a symmetric tensor, then

$$\sigma(t) \in k$$

roof: This follows from 4.7 since $\theta_{A\otimes\bar{k}/\bar{k}} = \theta_{A/k}\otimes 1$ by 4.5.

We leave to the reader to identify the multiplicities entioned in 4.9 with degrees of inseparability of suitable ield extensions.

At this level of specification we may entirely forget $_{B/A}$ and replace it by rationality statements. If we eep specializing we end up in Galois Theory as the following xample illustrates: Let k be a field and a an element of $\bar{k}$ hich is invariant under $\mathrm{Gal}(\bar{k}/k)$. If we apply 4.8 to finite subextension of degree m which contains a, e find that $(T-a)^m$ has coefficients in k and consequently $a^{p^n} \in k$ for some $n \in \mathbb{N}$ where p is the characteristic xponent of k.

Appendix. Remarks on the structure of $M_{n,\mathbb{Z}}^{(n)}$.

Let A be a commutative ring, V a free A-module of rank m , $n \in \mathbb{N}$. We will identify $\otimes_A^n \mathrm{End}_A(V)$ and $\mathrm{End}_A(\otimes_A^n V)$ and consider the natural action of S_n on $\otimes_A^n \mathrm{End}_A(V)$ and $\otimes_A^n V$. These are related by the formula

A.1 $\sigma(t(v)) = (\sigma t)(\sigma v)$ $\quad t \in \otimes_A^n \mathrm{End}_A V$, $v \in \otimes_A^n V$, $\sigma \in S_n$

from which it follows that

A.2 $\mathrm{End}_A(V)_A^{(n)}$ can be identified with the A-algebra of those A-linear endomorphisms which commute with the action of S_n .

In particular $\mathrm{End}_A(V)^{(n)}$ commutes with the anti-symmetry operator $\mathcal{a} = \sum_{\sigma \in S_n} \epsilon_\sigma \sigma$ and consequently $\mathrm{Ker}(\mathcal{a})$ is stable for $\mathrm{End}_A(V)^{(n)}$ which now operates on $\Lambda_A^n V$ in virtue of the exact sequence

$$0 \to \mathrm{Ker}(\mathcal{a}) \to \otimes_A^n V \to \Lambda_A^n V \to 0$$

We leave it to the reader to verify that $\mathrm{ld}_A^{(n)} : M_{n,A}^{(n)} \to$ coincides with this representation in case $V = A^n$.

As another consequence of A.2 and standard representat theory we get that $M_{n,\mathbb{Q}}^{(n)}$ is semi simple. $M_{n,\mathbb{F}_p}^{(n)}$ is semi simple (p rational prime) if and only if $p > n$.

The irreducible representations of $M_{n,\mathbb{C}}^{(n)}$ contained in $\otimes^n\mathbb{C}^n$ relates to those of $Gl_{n,\mathbb{C}}$ via Schur-theory. From the character table one gets

$$M_{2,\mathbb{C}}^{(2)} \cong \mathbb{C}\oplus M_{3,\mathbb{C}}$$

$$M_{3,\mathbb{C}}^{(3)} \cong \mathbb{C}\oplus M_{8,\mathbb{C}}\oplus M_{10,\mathbb{C}}$$

$$M_{4,\mathbb{C}}^{(4)} \cong \mathbb{C}\oplus M_{15}\oplus M_{20}\oplus M_{35}\oplus M_{45}$$

. . . . . In fact it can be shown that

$$1d : M_{n,\mathbb{C}}^{(n)} \to \mathbb{C}$$

is the only one dimensional representation of $M_{n,\mathbb{C}}^{(n)}$.

CHAPTER II

REPRESENTATION OF N-FOLD SECTIONS BY SYMMETRIC PRODUCTS.

II.1 The canonical subscheme of $C \times C^{(n)}$.

II.2 Geometry of the linear determinant.

II.3 n-fold sections.

II.4 Families of curves.

Appendix. Weil's theorem on symmetric functions.

II.1 The canonical subscheme of $C^{(n)} \times C$.

Let $C \to S$ be a flat morphism of schemes which satisfies

1.1. For all $s \in S$, any finite set of points of the s-fiber of $C \to S$ is contained in an open affine subset of C whose image is contained in an open affine of S .

Condition 1.1 is clearly stable under base change $T \to S$.

Let $C_S^n = C^n$ denote the product over S of n copies of C . The quotient of C^n under the action of the symmetric group will be denoted $C_S^{(n)}$ or $C^{(n)}$. $C_S^{(n)}$ is flat over S by I.1.1. If $T \to S$ is a base extension, then $C_T^{(n)}$ will denote $(C\times_S T)_T^{(n)}$ which is canonically isomorphic to $C_S^{(n)} \times_S T$ by I.1.2. Since C is flat over S we have the addition map (n,m) I.1.4.

1.2. $$C_S^{(n)} \times_S C_S^{(m)} \to C_S^{(n+m)}$$

The addition maps are affine morphisms as is easily seen and so is the map

1.3. $$C^{(n-1)} \times C \to C^{(n)} \times C$$

whose projection after $C^{(n)}$ is the addition map $(h-1,1)$ and projection after C is the second projection $C^{(n-1)}\times C \to C$. If we combine this with (I.1.5) we get

<u>Proposition 1.4</u>. $C^{(n-1)}\times C \to C^{(n)}\times C$ is a closed immersion.

This subscheme of $C^{(n)}\times C$ will be called the canonical subscheme.

II.2. Geometry of the linear determinant.

Let $f : X \to Y$ be a morphism of schemes which is finite, flat of rank n . If Y is an affine scheme we have constructed in I.4.3. a canonical section of the Y-scheme $X_Y^{(n)}$. By means of an affine open covering of Y is in this way constructed a canonical (Y) section

2.1. $$\theta_{X/Y} : Y \to X_Y^{(n)}$$

From I.4. 4-6 we deduce the following properties of $\theta_{X/Y}$:

2.2. $$X \xrightarrow[(\theta_{X/Y}\circ f,1)]{} X_Y^{(n)} \times_Y X$$

can be factored through the canonical subscheme 1.4

$$X_Y^{(n-1)} \times_Y X \to X_Y^{(n)} \times_Y X$$

2.3. If $$\begin{array}{ccc} X & \longrightarrow & Y \\ {\scriptstyle f}\downarrow & & \downarrow{\scriptstyle g} \\ Z & \longrightarrow & V \end{array}$$

is a commutative diagram in the category of schemes with X (resp. Z) finite, flat of rank n over Y (resp. V) then the following diagram is commutative

$$\begin{array}{ccc} Y & \xrightarrow{\ \theta_{X/Y}\ } & X_Y^{(n)} \\ {\scriptstyle g}\downarrow & & \downarrow{\scriptstyle f^{(n)}} \\ V & \xrightarrow{\ \theta_{Z/V}\ } & Z_V^{(n)} \end{array} \quad .$$

Let X be a scheme and let $\amalg^n X$ denote the disjoint union of n copies of X considered as a scheme over X in the natural way. Then we have the following commutative diagram

2.4.
$$\begin{array}{ccc} (\amalg^n X)_X^n & \longrightarrow & (\amalg^n X)_X^{(n)} \\ & {\scriptstyle s}\nwarrow & \uparrow{\scriptstyle \theta_{\amalg^n X/X}} \\ & & X \end{array}$$

where the horizontal morphism is the canonical projection and where s is the morphism whose projection after the ith factor is the ith inclusion $X \to \amalg^n X$.

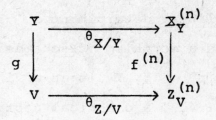

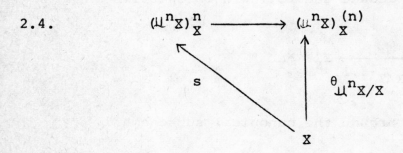

II.3. n-fold sections

Let $C \to S$ be a flat morphism of schemes which satisfies condition 1.1.

Definition 3.1. If $T \to S$ is a morphsim of schemes then an n-fold section of C over T is a closed subscheme F of $C \times_S T$ for which the projection $F \to T$ is finite flat of rank n. The set of n-fold sections of C over T is denoted $F^n_{C/S}(T)$.

By pull back of n-fold sections is in this way defined a contravariant functor $F^n_{C/T}$ from S-schemes to sets.

We will now define a natural transformation

3.2. $\varphi : F^n_{C/S} \to \mathrm{Hom}_S(\ ,C^{(n)}_S)$

as follows: Let F be an n-fold section of C over T, we have morphisms

$$F \to C \times_S T \ , \ p_1 : F \to C \ , \ p_2 : F \to T$$

and consequently the following sequence of maps

3.3. $$T \xrightarrow{\theta_{F/T}} F_T^{(n)} \xrightarrow{p_1^{(n)}} C_T^{(n)} \longrightarrow C_S^{(n)}$$

whose composit is $\varphi_T(F)$ by definition. One verifies easily that this defines a natural transformation.

<u>Theorem 3.4</u>. Let $C \to S$ be a flat morphism of schemes which satisfies condition 1.1. If the addition map $C_S^{(n-1)} \times_S C \to C_S^{(n)}$ is finite, flat of rank n, then the natural transformation

$$\varphi : F^n_{C/S} \to \mathrm{Hom}_S(\ ,C_S^{(n)})$$

is an isomorphism and the inverse transformation associates to $f : T \to C_S^{(n)}$ the pull back of $C^{(n-1)} \times_S C \to C_S^{(n)} \times C$ along f.

<u>Proof</u>. Assume that $C^{(n-1)} \times C \to C^{(n)}$ is finite, flat of rank n. Then the last sentence in the theorem defines a natural transformation

$$\psi : \mathrm{Hom}_S(\ ,C_S^{(n)}) \to F^n_{C/S}$$

Let us first show that $\psi \circ \varphi = \mathrm{id}$:

Let F be an n-fold section of C over T and consider the diagram

$$\begin{array}{ccc} C\times T & \xrightarrow{1\times \varphi_T(F)} & C\times C^{(n)} \\ \uparrow & & \uparrow \\ F & & C\times C^{(n-1)} \end{array}$$

We want to show that the restriction of $\varphi_T(F)$ to F can be factored through $C\times C^{(n-1)}$. The restriction of $1\times\varphi_T(F)$ to F is the composite of the map $F \to F\times_T F^{(n)}$ described in 2.2 and the map $F\times_T F_T^{(n)} \to C\times_S C_S^{(n)}$ induced by the projection $F \to C$. Consider the following commutative diagram

$$\begin{array}{ccccc} F & \longrightarrow & F\times_T F_T^{(n)} & \longrightarrow & C\times_S C_S^{(n)} \\ & \searrow & \uparrow & & \uparrow \\ & & F\times_T F_T^{(n-1)} & \longrightarrow & C\times_S C_S^{(n-1)} \end{array}$$

where the triangle is the one discussed in 2.2. It follows that the pull back F' of $C\times C^{(n-1)}$ along $1\times\varphi_T(F)$ contains F as a closed subscheme. However F and F' are both finite, flat of rank n over T. Consequently $F = F'$.

Proof of $\varphi\cdot\psi = \mathrm{id}$:

Let $f : I \to C_S^{(n)}$ be a S-morphism and let F denote the pull vack of $D = C\times_S C_S^{(n-1)}$ along f. The problem is to prove that $\varphi_T(F) = f$. We have $\varphi_T(F) = \varphi_{C^{(n)}}(D)\cdot f$ as it follows from the more general

<u>Lemma 3.5</u>. Let $f : T \to V$ be an S-morphism. If $E\in F^n_{C/S}(V)$ and $F = F^n_{C/S}(f)(E)$ then $\varphi_T(F) = \varphi_V(E)\cdot f$.

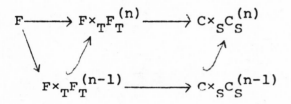

The lemma is a consequence of 2.3.

We have now reduced to prove $\varphi_{C^{(n)}}(D) = \mathrm{id}$. Let π_n denote the projection $C^n \to C^{(n)}$. It suffices to show $\varphi_{C^{(n)}}(D)\cdot\pi = \pi$. Consider the commutative diagram

$$\begin{array}{ccccc} C^n & \xrightarrow{\pi_n} & C^{(n)} & & \\ \uparrow & & \uparrow & & \\ \coprod^n C^n & \xrightarrow{g} & C\times C^{(n-1)} & \xrightarrow{p_1} & C \end{array}$$

The restriction of g to the ith component of $\coprod^k C^k$ is $1\times\pi_{n-1}$. From the above diagram results the following commutative diagram

$$\begin{array}{ccccc} C^n & \longrightarrow & C^{(n)} & & \\ \downarrow{\scriptstyle\theta} & & \downarrow{\scriptstyle\theta} & & \\ (\coprod^n C^n)^{(n)}_{C^n} & \longrightarrow & (C\times C^{(n-1)})^{(n)}_{C^{(n)}} & \longrightarrow & C_S^{(n)} \end{array}$$

It is now easy to finish the proof by means of 2.4.

Q.E.D.

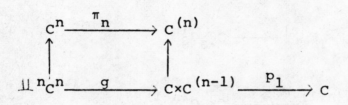

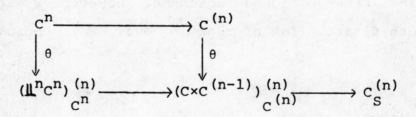

II.4. Families of curves.

The aim of this section is to prove the assumption made in 3.4 for a smooth family of curves.

Proposition 4.1. Let S be a locally noetherian scheme, $C \to S$ a smooth morphism of schemes whose fibers are irreducible curves. Then the addition map

$$C^{(n-1)} \times_S C \to C^{(n)}$$

is finite, flat of rank n .

Proof. $C^{(n-1)} \times_S C \to C^{(n)}$ is affine. Passing to the case where S is affine and noetherian and C affine one sees that $C_S^{(n-1)} \times_S C \to C_S^{(n)}$ is finite. $C^{(n-1)} \times C$ and $C^{(n)}$ are both flat over S ; one reduces now the question to the case where S is a field by a standard technique [S.G.A.], 60-61, Cor. 5.9 de Th. 5.6. Let $S = \operatorname{spec} k$ where k is a field. $C_k^{(n-1)} \times_k C$ and $C_k^{(n)}$ are both non singular and irreducible and the map $C_k^{(n-1)} \times_k C \to C_k^{(n)}$ is finite and generally surjective. Flatness is now a consequence of a well known lemma, see [E.G.A.] Chap. 0, 17.3.5. -- A look at the generique fiber of $C^{(n-1)} \times C \to C^{(n)}$ will show that the degree of the map is n .

Q.E.D.

Appendix. Weil's theorem on symmetric functions.

We take a closer look at the situation discussed in Chap. II in case the base scheme is a field by discussing "The fundamental theorem on symmetric functions" [W] Th. 1.

Let X be a scheme of finite type over the field k. For a field extension $k \to l$ let $\mathrm{Cyc}^n_{X/k}(l)$ denote the set of positive 0-cycles of degree n (considered as a subset of the free group generated by the closed points of $X \times_k l$). To an n-fold section F of X over l is in an obvious way associated an element $\mathrm{div}\,(F) \in \mathrm{Cyc}^n_{X/k}(l)$ and we get a functorial map

$$F_{X/k}(l) \to \mathrm{Cyc}^n_{X/k}(l)$$

Proposition. Let F_1 and F_2 be two n-fold sections of X over l. Then, $\mathrm{div}\,(F_1) = \mathrm{div}\,(F_2)$ if and only if the corresponding l-rational points 3.2 of $X_k^{(n)}$ are the same.

Proof. Pass to the algebraic closure of l and apply I.4.7.

This suggests the existence of a map

$$\mathrm{Cyc}^n_{X/k}(l) \to X_k^{(n)}(l)$$

such that the following diagram is commutative

$$\begin{array}{ccc} F^n_{X/k}(l) & \longrightarrow & Cyc^n_{X/k}(l) \\ \downarrow & \swarrow & \\ X_k^{(n)}(l) & & \end{array}$$

such a map is constructed as follows. Let $D = \Sigma n_P P \in Cyc^n_{X/k}(l)$, with $dg(P) = d_P$ and let φ_P denote the l-rational point of $X_k^{(d_P)}$ corresponding to the d_P fold section P. $\varphi_D \in X_k^{(n)}(l)$ are now to be constructed from the φ_P's and the k_P's by means of the addition maps 1.2.

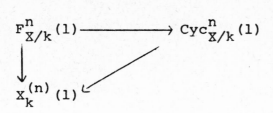

CHAPTER III

INVERTIBLE SHEAVES AND RATIONAL MAPS INTO $C^{(g)}$

III.1 The canonical invertible sheaf "parametrized" by $C^{(g)}$.

III.2 The rational map into $C^{(g)}$ defined by an invertible sheaf.

III.3 Cohomology of $\mathcal{L} \otimes D$.

III.4 Dominant rational maps into $C^{(g)}$.

III.1. The canonical invertible sheaf "parametrized" by $C^{(g)}$.

Throughout Chapter III $C \to S$ denotes a flat, proper morphism of schemes whose fibers are geometrically reduced and geometrically irreducible curves of genus g , i.e.

$$\dim_{k(s)} H^1(C_s, \mathcal{O}_{C_s}) = g$$

for all points $s \in S$. It is furthermore assumed that $C \to S$ satisfies condition II.1.1 and that S is locally noetherian. All S-schemes considered are assumed to be of finite type over S .

Notation. Whenever $T \to S$ is a morphism of schemes, an invertible sheaf on $C_T = C \times_S T$ will be called an invertible sheaf on C parametrized by T . An open subscheme U of T will be called S-dense if U is universally schematically dense relative to S in the terminology of Grothendieck [E.G.A.] IV. 20.2.1. By an invertible sheaf on C "parametrized" by T is understood an equivalence class of pairs (U,L) where U is an S-dense open of T and L is an invertible sheaf on $C_U = C \times_S U$, (U_1, L_1) and (U_2, L_2) are equivalent if there exists an S-dense open $W \subseteq U_1 \cap U_2$ such that L_1 and L_2 has the same restriction to C_W .

Let X and Y be S-schemes. By an S-rational map from X to Y is understood a pseudo-morphism from X to Y relative to S [E.G.A.] IV. 20.5.1.

<u>Proposition 1.1.</u> Let J denote the sheaf of ideals for the closed subscheme

$$C_S^{(g-1)} \times_S C \to C_S^{(g)} \times_S C \ .$$

There exists an S-dense open subscheme U of $C^{(g)}$ such that the restriction of J to $C_U = C\times_S U$ is an invertible sheaf.

If $C \to S$ is smooth then we may take $U = C^{(g)}$.

<u>Proof</u>. Let D denote the open subscheme of C consisting of those points of C where $C \to S$ is smooth. Since $C \to S$ is flat it follows from the Jacobian criterion that the fiber of $D \to S$ at a point $s \in S$ is the smooth locus of C_s . C_s is smooth at the generic point so D is S-dense in C by [E.G.A.] IV. 11.10.10 and consequently $U = D_S^{(g)}$ is an S-dense open subscheme of $C_S^{(g)}$.

Next we want to prove that the following diagram is cartesian.

$$\begin{array}{ccc} C\times D^{(g)} & \longrightarrow & C\times C^{(g)} \\ \uparrow & & \uparrow \\ D\times D^{(g-1)} & \longrightarrow & C\times C^{(g-1)} \end{array}$$

Since the two horizontal arrows are open immersions we may verify this on each fiber separately. So we may assume that S is a field and even an algebraically closed one. Now the statement is easily verified by considering the k-rational points.

We want to prove that the sheaf of ideals defining the closed immersion

$$D \times D^{(g-1)} \to C \times D^{(g)}$$

is locally generated by a non zero divisor. But this closed immersion can be factored through the open immersion

$$D \times D^{(g)} \to C \times D^{(g)}$$

Consequently, we are reduced to the case where $C \to S$ is smooth.

In case where S is a field the proposition is clear from the fact that $C \times C^{(g-1)}$ is an irreducible closed subvariety of the non singular variety $C \times C^{(g)}$ of codimension 1 .

Now assume S is an arbitrary locally noetherian scheme (and $C \to S$ is smooth). To get a local equation for $C^{(g-1)} \times C$ at a point x of $C^{(g)} \times C$ we just take any element α of $\mathcal{O}_{C^{(g)} \times C, x}$ whose restriction to the fiber of $\pi : C \times C^{(g)} \to S$ at $s = \pi(x)$ is a local equation for the fiber of $C^{(g-1)}\times C \to S$ at s . It is now easy to conclude by Nakayamas lemma, taking into account that $C^{(g-1)}\times C$ is flat over S , that α is a local equation for $C^{(g-1)}\times C$ and that α is a non zero divisor.

Q.E.D.

Definition 1.2. With the notation of proposition 1.1, (J^{-1},U) will be called the canonical invertible sheaf "parametrized" by $C^{(g)}$ and will be denoted $\mathcal{L}$. The restriction of $C\times C^{g-1}$ to U will be called the canonical g-fold section of C "over" $C^{(g)}$.

III.2. The rational map into $C^{(g)}$ defined by an invertible sheaf.

An invertible sheaf L on $C_T = C\times_S T$ where T is an S-scheme is said to be of degree n if $dg_t(L) = n$ for all $t\in T$, where $dg_t(L) = h^0(C_t,L_t) - h^1(C_t,L_t) + 1 - p_a(C_t)$. Recall that $dg_t(L)$ is additive in L (Riemann-Roch) and locally constant on T for fixed L.

Proposition 2.1. Let L be an invertible sheaf on C parametrized by T which satisfies:

2.2. There exists an S-dense open U of T such that $H^1(C_t,L_t) = 0$ for all $t\in U$,

then there exists an S-dense open V of T and a g-fold section F of C over V whose sheaf of ideals J is invertible on $C\times_S V$ and such that

$$J^{-1} \cong L|_V \otimes p_2{}^*E$$

for some invertible sheaf E on V, where $p_2 : C\times_S V \to V$ denotes the second projection.

The pair (V,F) is unique in the sense that if (V_i,F_i), $i = 1,2$, has the above property, then F_1 and F_2 has the same restriction to $V_1\cap V_2$.

<u>Proof</u>. We make free use of the basic facts about cohomology of coherent sheaves as they are exposed in [E.G.A.]III §7.

It is a consequence of 2.2 that $p_{2*}L$ is locally free of rank 1 and that the restriction map

2.3. $$(p_{2*}L)\otimes_{\mathcal{O}_T} k(t) \to H^0(C_t, L_t) \quad \text{is an isomorphism}$$ for all $t \in U$

Let $E = (p_{2*}L)^{-1}$. The canonical map $(p_{2*}L)\otimes E \to p_{2*}(L\otimes p_2{}^*E)$ is an isomorphism since E is locally free. Replacing L by $L\otimes p_2{}^*E$ we may in addition to 2.2 and 2.3 assume

2.4. $$p_{2*}L \cong \mathcal{O}_U$$

2.3 and 2.4 implies that there exists a global section s of L over $C_U = C\times_S U$ whose restriction to $H^0(C_t, L_t)$ is non zero for all $t \in U$ and that such a section is unique up to multiplication with a global unit of $\mathcal{O}_U$.

s defines a map $\mathcal{O}_U \to L$. We want to prove that this map is injective and that the closed subscheme F of C_U defined by the resulting inclusion $L^{-1} \hookrightarrow \mathcal{O}_{D_U}$ is flat over U. The problem is local so let x be a point of C_U, $t = p_2(x)$.

Let the stalk at x of the dual map of $\mathcal{O}_{C_U} \xrightarrow{s} L$, $L_x^{-1} \to \mathcal{O}_{C_U,x}$ be given by the element α of $\mathcal{O}_{C_U,x}$.

By the fact that the fibers of $C_U \to U$ are reduced and irreducible and the fact that s has a non zero restriction to C_t follows that the image of α by the restriction map $\mathcal{O}_{C_U,x} \to \mathcal{O}_{C_t,x}$ is non zero. It follows from [S.G.A.] 60-61 Chap. IV Cor. 5.7 de Th. 5.6 that α is a non zero divisor and that $\mathcal{O}_{C_U,x}/(\alpha)$ is flat over $\mathcal{O}_{U,t}$.

The fibers of $F \to U$ are finite and $F \to U$ is proper. So by a theorem of Chevalley-Grothendieck [E.G.A.] III 4.4.2 we conclude that $F \to U$ is affine. Furthermore we know that $F \to U$ is finite and flat, and we get that it has constant rank g by considering the long cohomology sequence of $(t \in U)$

$$0 \to \mathcal{L}_t^{-1} \to \mathcal{O}_{C_t} \to \mathcal{O}_{F_t} \to 0$$

Uniqueness: Let F_1 and F_2 be g-fold sections of C over U which satisfies the requirement of Proposition 2.1. Then there exists invertible sheaves E_1 and E_2 on U and exact sequences, $i = 1,2$, $L_i = L \otimes p_2{}^*E$

$$0 \to L_i^{-1} \to \mathcal{O}_{C_U} \to \mathcal{O}_{F_i} \to .$$

which implies that $p_{2*}L_i \approx \mathcal{O}_U$. Now $p_{2*}L_i = E_i \otimes p_{2*}L$

so $E_1 \approx E_2$. Let the two sequences above correspond to sections s_1 and s_2 of L . Then as above, s_1 and s_2 differs only by a global unit of $\mathcal{O}_U$, hence $F_1 = F_2$.

Q.E.D.

Remark 2.5. With the notation of 2.1 assume $T \to S$ is is flat. Then condition 2.2 can be replaced by

2.6. $H^1(C_t, L_t) = 0$ whenever t is a generic point of an irreducible component or embedded component of the fiber of $T \to S$ which contains t

as it follows from the upper semi-continuity theorem and [E,G,A,] IV. 11.10.10.

III.3. Cohomology of $\mathcal{L}\otimes D$.

Proposition 3.1. Let $\mathcal{L}$ denote the canonical invertible sheaf "parametrized" by $C^{(g)}$ (1.2) , and let D be an invertible sheaf of degree 0 "parametrized" by S . The pull back of D to $C^{(g)}$ will still be denoted D . Then $\mathcal{L}\otimes D$ has degree g and

$$H^1(C_t,(\mathcal{L}\otimes D)_t) = 0$$

for all t in an S-dense open of $C_S^{(g)}$.

Proof. According to remark 2.5 we may assume S = Spec (k) where k is a field and that t is the generic point of $C_S^{(g)} = C_k^{(g)}$.

Let K denote the function field of C and let $K^{(g)}$ resp. K^g denote the function field of $C_k^{(g)}$ resp. C^g . We have two canonical maps

$$\mathrm{Spec}\ K^g \to \mathrm{Spec}\ K^{(g)} \to C^{(g)}$$

The pull back to K^g of the canonical invertible sheaf "parametrized" by $C^{(g)}$ is easily seen to be the dual sheaf of the sheaf of ideals defining the closed subscheme $P_1,\ldots,P_g$ of $C\times_k K^g$ where P_i is the K^g-rational point

$$\mathrm{Spec}\ (K^g) \xrightarrow{p_i} \mathrm{Spec}\ K \to C$$

where p_i is the map induced on the generic points by the i'th projection $C^g \to C$.

According to the theorem of Riemann-Roch we have $h^1(D) \leq g$. The proposition now follows from the following lemma

<u>Lemma 3.2</u>. (Weil-Rosenlicht). Let C/k be a geometrically reduced and geometrically irreducible complete curve, $k \to K$ a field extension, P a K-rational of C which lies over the generic point of C , L_P the invertible sheaf on $C\times_k K$ defined by

3.3. $$0 \to L_P^{-1} \to \mathcal{O}_{C_K} \to K(P) \to 0$$

and D an invertible sheaf on C . Then

$$h^1(C_K, L_P \otimes D) = \sup\{0, h^1(C,D)-1\}$$

<u>Proof</u>. $\bar{C} = C\otimes_k K$, $D_k = D\otimes_k K$, ω denotes the dualizing sheaf for C , $\omega_k = \omega\otimes_k K$, see [M_1]p. 79 . ω is torsion free of rank 1, loc. cit. From the exact sequence 3.3 we get the following sequence

$$0 \to \omega_k \otimes D_k^{-1} \otimes L_P^{-1} \to \omega_k \otimes D_k^{-1}$$
$$\to \omega_k \otimes D_k^{-1} \otimes K(P) \to 0$$

which is exact since

$$\mathcal{T}or_1(k(P), D_k^{-1}\otimes\omega_k) = \mathrm{Tor}_1^{\mathcal{O}}(k(P), (D_k^{-1}\otimes\omega_k)_P) \text{ where } \mathcal{O} = \mathcal{O}_{\bar{C},P}$$

which is equal to zero since ω_k is torsion free and P is a simple point of $\bar{C}$.

We get an exact sequence

$$0 \to H^0(\bar{C}, \omega_k\otimes D_k^{-1}\otimes L_P^{-1}) \to$$

$$H^0(\bar{C}, \omega_k\otimes D_k^{-1}) \to \omega_k\otimes D_k^{-1}\otimes k(P)$$

Suppose $H^1(C,D) \cong H^0(C, \omega\otimes D^{-1}) \neq 0$, then the last map in the sequence is surjective since $H^0(\bar{C}, \omega_k\otimes D_k^{-1}) = H^0(C, \omega\otimes D^{-1})\otimes_k K$ and any non zero global section of $\omega\otimes D^{-1}$ has a non zero image in $\omega_k\otimes D_k^{-1}\otimes k(P)$. Hence by duality

$$h^1(\bar{C}, D_k\otimes L_P) = h^0(\bar{C}, \omega_k\otimes D_k^{-1}\otimes L_P^{-1})$$

$$= h^0(\bar{C}, \omega_k\otimes D_k^{-1})-1 = h^1(C,D)-1$$

Q.E.D.

III.4. Dominant rational maps into $C^{(g)}$

Definition 4.1. Let $f : X \to Y$ be an S-morphism. f is called S-dominant if for all $T \to S$ the inverse image of a T-dense open of $Y \times_S T$ by $f \times 1$ is an T-dense open of $X \times_S T$. If f is an S-rational map, then f is called S-dominant if there exists an S-dense open U of X on which f is defined such that $U \to Y$ is an S-dominant morphism.

If $f : X \to Y$ is an S-dominant S-rational map and L is an invertible sheaf on C "parametrized" by Y then it is clear that we may consider f^*L , the pull back of L along f .

Remark 4.2. If $f : X \to Y$ is an S-rational map and X is flat and of finite type over S then in order for f to be S-dominant it suffices that for all $s \in S$ the image of a generic point of an irreducible component (incl. embedded components) of X_s be a generic point of an irreducible component (incl. embedded components) of Y_s as it follows from [E.G.A.] IV. 11. 10. 10.

Proposition 4.3. Let T be an S-scheme, locally of finite type and

$$f_1, f_2 : T \to C_S^{(g)}$$

two S-dominant S-rational maps. If $f_1^*\mathcal{L} \cong f_2^*\mathcal{L}\otimes\pi^*E$ where E is an invertible sheaf on some S-dense open $T(\pi : C\times T \to T$ denotes the projection) then $f_1 = f_2$.

Proof. Shrinking T we may assume that f_1 and f_2 are morphisms and that

$$H^1(C_t,(f_i^*\mathcal{L})_t) = 0$$

for all $t\in T$ by 3.1. It follows from 2.1 that the canonical g-fold section of C "over" $C^{(g)}$ has the same pull back along f_1 and f_2 . We can now conclude $f_1 = f_2$ by the following lemma.

Lemma 4.4. Let $f : T \to C^{(g)}$ be an S-dominant S-rational map and let F denote the pull back of the canonical g-fold section of C "over" $C^{(g)}$. Then f is the rational map defined by F in the sense of II.3.2.

Proof. Let D denote the open set of C consisting of the points where $C \to S$ is smooth. Shrinking T we may assume that f is a morphism and that it takes values in $D^{(g)}$. The lemma now follows from the cartesian diagram established in the proof of 1.1 and II Th. 3.4.

Theorem 4.5. With the notation of 3.1 there exists a unique

S-dominant S-rational map $r_D : C^{(g)} \to C^{(g)}$ such that

4.6. $$r_D{}^*\mathcal{L} = \mathcal{L}\otimes D\otimes \pi^* E$$

where E is an invertible sheaf and an open S-dense subset of $C^{(g)}$ and π denotes the second projection $C\times C^{(g)} \to C^{(g)}$.

Proof. The uniqueness is clear from 4.3. The conjugation of 2.1 and 3.1 defines a rational map $r_D : C^{(g)} \to C^{(g)}$. If we can prove that r_D is dominant it follows from 4.4 that r_D satisfies condition 4.6 above. To prove that r_D is dominant we may according to 4.2 assume that the base scheme is the spectrum of an algebraically closed field k and we may consider r_D as a morphism from $k^{(g)}$ to $C^{(g)}$. In this case the dominance of r_D is proved by Weil [W] in the singular case and Rosenlicht [R_2] in the singular case. We find it reasonable to reproduce a part of Rosenlicht's proof. The notation will be the one introduced in the proof of 3.1.

Let s be the non zero global section of $\mathcal{L}\otimes D$. Suppose

4.7. $s(P) \neq 0$, whenever P is a singular point of $C\times_k K^g$

We will finish the proof under the assumption 4.7 and then prove 4.7 below. Then it follows from the cartesian diagram established in the proof of 1.1 that r_D takes value in $C^{(g)}_{n.s.}$, it is now an easy consequence of II.3.4 that $r_D^*\mathcal{L} = \mathcal{L}\otimes D$ ($\mathcal{L}$ is defined over $C^{(g)}_{n.s.}$ according to the proof of 1.1). If r_D is not dominant then it follows that $\mathcal{L}\otimes D$ and consequently $\mathcal{L}$ can be descended to $C\times_k L$ where L is a subextension of $K^{(g)}$ of transcendence degree $\leq g-1$. This implies according to II.3.5. that the morphism from $K^{(g)}$ to $C^{(g)}$ defined by $\mathcal{L}$ can be factored through

$$\operatorname{Spec} K^{(g)} \to \operatorname{Spec} L$$

which is impossible since the morphism defined by $\mathcal{L}$ is the canonical inclusion by Lemma 4.4.

Proof of 4.7. (After [R_2] p. 515-516)

The proof is in three parts

1° Let $\pi : C \to E$ be a birational morphism of C onto a complete curve E. If 4.7 is true for E then it is true for C.

2° 4.7 is true for $\tilde{C}_{\mathfrak{m}}$ where $\tilde{C}$ denotes the normalization of C, $\mathfrak{m}$ a positive divisor on $\tilde{C}$ and $\tilde{C}_{\mathfrak{m}}$ the Rosenlicht curve of module $\mathfrak{m}$.

3° There exists a birational morphism of C onto a Rosenlicht curve of suitable module.

Proof of 1° : We may assume that D is non trivial. Let F be an invertible sheaf on E whose pull back to C is D. Such a sheaf is well known to exist. Let C resp. E have arithmetic genus c resp. e and let L denote the function field of C^e. $P_1, \ldots, P_e$ denote the e canonical rational points of $C \times_k L$, $\pi(P_i) = Q_i$. L_i denotes the invertible sheaf on $E \times_k L$ defined by

4.8. $$0 \to L_i^{-1} \to \mathcal{O}_{E\times L} \to L(Q_i) \to 0$$

$\pi^* L_i$ will still be denoted L_i. A point R on a curve is said to be a base point for an invertible sheaf G if the stalk of G at R is not generated by the global sections of G. Let P be a singular point of C which is a base point for $DL_1L_2 \ldots L_c$, we want to prove that $\pi(P) = Q$ is a base point for $FL_1L_2 \ldots L_e$. Since $\Gamma(E, FL_1 \ldots L_e) \subseteq \Gamma(C, DL_1 \ldots L_e)$ and $L(P) = L(G)$ it suffices to prove that P is a base point for $DL_1 \ldots L_e$. P and D are rational over k consequently P is also a base point for $D_i = DL_1 \ldots L_{e-1}L_i$ $i \geq c$. We have $h^0(C, D_i) = 1$ and $h^0(DL_1 \ldots L_{c-1}) = 0$ by Lemma 3.2, Riemann-Roch and the fact that D is non trivial. Tensoring 4.8 by D_i it follows that P_i is not a base point for D_i. Let s_i be the non zero global section of D_i; 4.8 defines inclusions

$$\theta_i : D_i \to DL_1 \ldots L_e$$

To prove that P is a base point for $DL_1 \ldots L_e$ it suffices to prove that $\theta_i(s_i)_{i=c,c+1,\ldots,e}$ generates $H^0(C,DL_1\ldots L_e)$. Now $h^0(C,CL_1\ldots L_e) = e-c+1$ as it follows from 3.2 and Riemann-Roch so it suffices to prove that $\theta_i(s_i)_{i=c,\ldots,e}$ are linearly independent. Repeated application of 4.8 shows that for $i,j \geq c$ we have $\theta_i \otimes L(P_j) = 0$ if and only if $i \neq j$ which concludes the proof of 1° since P_i is not a base point for D_i.

<u>Proof of 2°</u> : For the following facts as well as for a proof of 3° we refer the reader to $[R_1]$ or [S] p. 68-71. We may assume that C is non singular of genus c. Let the positive divisor $\mathfrak{m}$ have degree $m \geq 2$, then $C_{\mathfrak{m}} = E$ has arithmetic genus $e = c+m-1$. E had precisely one singular point P and $\mathcal{O}_C \otimes_{\mathcal{O}_E} k(P) = \mathcal{O}_C/\mathfrak{m}$. Let $\pi : C \to E$ denote the canonical map and consider the exact sequence

$$0 \to \mathfrak{m}^{-1}\pi^*(DL_1\ldots L_e) \to \pi^*(DL_1\ldots L_e)$$

$$\to \pi^*(DL_1\ldots L_e)\otimes \mathcal{O}_C/\mathfrak{m} \to 0$$

If P were a fixed point for $DL_1 \ldots L_e$ then a non zero global section of $DL_1 \ldots L_e$ would induce a non zero global section of $\mathfrak{m}^{-1}\pi^*(DL_1 \ldots L_e)$, but we are going to prove that $H^0(C, \mathfrak{m}^{-1}\pi^*(DL_1 \ldots L_e) = 0$. First note that $\mathfrak{m}^{-1}\pi^*(DL_1 \ldots L_m)$ is non trivial by an argument similar to the one immediately preceeding the beginning of the proof of 4.7. Hence,
$h^1(C, \mathfrak{m}^{-1}\pi^*(DL_1 \ldots L_m)) = c-1$ and by Lemma 4.4
$h^1(C, \mathfrak{m}^{-1}\pi^*(DL_1 \ldots L_e) = 0$; finally, Riemann-Roch gives

$$h^0(C, \mathfrak{m}^{-1}\pi^*(DL_1 \ldots L_e)) = -m + e + 1 - c = 0$$

Q.E.D.

Remark 4.9. Let X and Y be S-schemes. According to definition 4.1, the projection $X \times_S Y \to X$ is S-dominant.

Corollary 4.10. Let T be an S-scheme and D an invertible sheaf of degree zero on C "parametrized" by T and let $\mathcal{L} \otimes D$ denote the invertible sheaf on C "parametrized" by $C^{(g)} \times_S T$ which is the tensor product of the pull back of $\mathcal{L}$ resp. D along the 1. resp. 2. projection. Then there exist a unique S-dominant S-rational map

$$r_D : C^{(g)} \times_S T \to C^{(g)}$$

such that

$$r_D{}^*\mathcal{L} = \mathcal{L}\otimes D\otimes\pi^*E$$

where E is an invertible sheaf on an S-dense open of $C^{(g)}\times T$ and $\pi : C^{(g)}\times T\times C \to C^{(g)}\times T$ denotes the projection.

Proof. Replace T by an S-dense open U of T on which D is defined, then make the base extension $U \to S$ and apply 4.5.

Q.E.D.

Remark 4.11. With the notation of 4.10 the S-rational map

$$(r_D, p_2) : C^{(g)}\times_S T \to C^{(g)}\times_S T$$

is S-dominant as follows from the proof above.

CHAPTER IV

CONSTRUCTION OF THE PICARD SCHEME
OF A FAMILY OF CURVES

IV.1 Seesaw principle, revised.

IV.2 Relative Picard functor, notation, theorem.

IV.3 Construction of a group scheme.

IV.4 Proof of the theorem on representability.

Appendix. The closed immersion of a family of curves into its Picard scheme.

IV.1 Seesaw principle, revised.

In this section we prove a version of the seesaw principle which in case of a reduced and irreducible base scheme is identical with the one given in [M].

We make free use of the basic theorem on base change of cohomology as exposed in [E.G.A.]III,§7 or in [M]; but first a preparation:

Lemma 1.1. Let S be a locally noetherian scheme, U a schematically dense open subscheme of S and F a coherent sheaf on S whose restriction to U is locally free. If $\dim_{k(s)} F \otimes k(s)$ is locally constant on U, then F is locally free on all of S.

Proof. We may assume that $\dim F \otimes k(s)$ is constant $= n$. We can now cover S with open affines U such that $F|_U$ is generated by n global sections. Consequently we are reduced to prove the following: Let A be a noetherian ring, F an A-module and $f : A^n \to F$ a surjective A-linear map. If 1) $f \otimes A/m$ is an isomorphism for all maximal ideals m of A and 2) there exist elements $s_1, \ldots, s_p$ of A such that $A \to \pi A_{s_i} = B$ is injective and $\pi f \otimes_A A_{s_i} = f \otimes_A B$ is an isomorphism, then f is an isomorphism. Noting that B is flat over A this statement follows simply from the following commutative, exact diagram

$$\begin{array}{ccccccccc}
 & & & & 0 & & & & \\
 & & & & \downarrow & & & & \\
0 & \rightarrow & K & \rightarrow & A^n & \rightarrow & F & \rightarrow & 0 \\
 & & \downarrow & & \downarrow & & \downarrow & & \\
0 & \rightarrow & K\otimes_A B & \rightarrow & A^n\otimes B & \rightarrow & F\otimes_A B & \rightarrow & 0
\end{array}$$

Q.E.D.

Proposition 1.2. Let S be a locally noetherian scheme, $f : X \rightarrow S$ a flat and proper map whose geometric fibers are reduced and irreducible, L an invertible sheaf on X and U a schematically dense open subscheme of S. If the restriction of L to $f^{-1}(U)$ is of the form f^*L_U where L_U is an invertible sheaf on U, then L is of the form f^*L_S where L_S is an invertible sheaf on S.

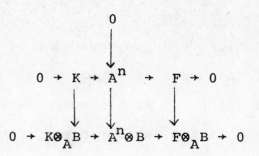

Proof. For $s \in S$ let L_s denote the restriction of L to X_s, the fiber of f at s. L_s is trivial if and only if $H^0(X_s, L_s) \neq 0$ and $H^0(X_s, L_s^{-1}) \neq 0$, see [M]. If follows from the upper semicontinuity theorem that the set of $s \in S$ such that L_s is trivial is a closed subset of S; by assumption this set contains U, so L_s is trivial for all $s \in S$. To prove the proposition we may assume $S = \operatorname{Spec} A$ where A ia a noetherian ring. As a consequence of Grothendieck's theorem on coherens of cohomology sheaves we can find a finite complex $C^\bullet$ ($C^n = 0$ for $n < 0$) of locally free, finitely generated A-modules which is homotopic to a Čech complex associated with f and L. Let T(resp.I) denote the cokernel (resp. image) of $C^0 \to C^1$. For $s \in S$ we have

$$\dim T \otimes k(s) = \dim C^0 \otimes k(s) - \dim C^1 \otimes k(s) + \dim H^0(X_s, L_s)$$

Since $\dim H^0(X_s, L_s) = 1$ it follows that $s \mapsto \dim T \otimes k(s)$ is locally constant on S. The next step is to prove that T is locally free on U. For this consider the following commutative, exact diagram

$$\begin{array}{ccccccccc}
 & & & & & & 0 & & \\
 & & & & & & \downarrow & & \\
 & & & & & & \mathrm{Tor}_1^A(T, k(s)) & & \\
 & & & & & & \downarrow & & \\
 & & H^0(L) \otimes k(s) & \to & C^0 \otimes k(s) & \to & I \otimes k(s) & \to & 0 \\
 & & \downarrow r_s & & \downarrow & & \downarrow & & \\
0 & \to & H^0(X_s, L_s) & \to & C^0 \otimes k(s) & \to & C^1 \otimes k(s) & &
\end{array}$$

By assumption we have r_s surjective for all $s \in U$ so by snake lemma $\mathrm{Tor}_1^A(T, k(s)) = 0$ for all $s \in U$ i.e. by a well known lemma T is locally free on U. From Lemma 1.1 we conclude that T is locally free on all of S which implies that $H^0(L) = f_*L$ is locally free on S and that $r_s : (f_*L) \otimes k(s) \to H^0(X_s, L_s)$ is an isomorphism for all $s \in S$. Replacing L by $L \otimes f^*f_*L^{-1}$ we may now assume that $f_*L = \mathcal{O}_S$. It follows from standard facts about flatness [S.G.A.] 60-61, Chap IV that L is trivial.

Q.E.D.

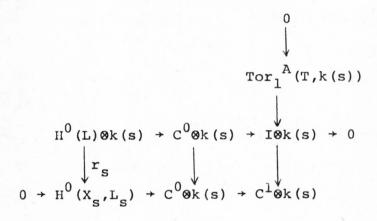

Remark 1.4. For later use note that the following fact (which is a little hard to dig out of [E.G.A.] III, §7) can be read off diagram 1.3: Let S and $F : X \to S$ be as in 1.2., F a locally free coherent sheaf on X and $s \in S$. If

$$(f_*F) \otimes k(s) \to H^0(X_s, F_s)$$

is surjective then f_*F is free in a neighborhood of s and $(f_*F) \otimes k(t) \to H^0(X_t, F_t)$ is an isomorphism for all t in a neighborhood of s.

.2. Relative Picard functor, notation, theorem.

Throughout the rest of Chapter IV $f : C \to S$ notes a flat and proper morphism of schemes whose ometric fibers are reduced and irreducible curves of ithmetic germs $g \geq 1$. S is assumed to be noetherian d f is assumed to satisfy condition II.1.1.

All S-schemes considered are assumed to be of nite type over S. A-morphism $S' \to S$ is called an fp-extension if $S' \to S$ is faithfully flat and of finite esentation (= finite type in this case).

On the category of schemes over S we will put the fp-topology in the sense of Grothendieck. For an teger n we have the contravariant functor

$$'\mathrm{Pic}^n_{C/S} : \mathrm{Sch}/S \Rightarrow \mathrm{Sets}$$

ic naif", whose T-valued points are classes of invertible eaves on $C_T = C \times_S T$ of degree n (cf. the first w lines of III.2.), two invertible sheaves L_1 and L_2 long to the same class if and only if $L_1 L_2{}^{-1} = p_2^* L$ for me invertible sheaf L on T. The fffp-sheaf associated th $'\mathrm{Pic}_{C/S}$ will be denoted by

$$\mathrm{Pic}^n_{C/S} : \mathrm{Sch}/S \Rightarrow \mathrm{Sets}$$

just by Pic^n when no confusion is possible.

From Grothendieck's [T.D.T.E.] V, n°2 we get the following

basic facts:

2.1. If T is an S-scheme then $'\mathrm{Pic}^n_{C/S}(T) \to \mathrm{Pic}^n_{C/S}(T)$ is injective.

2.2. The above map is bijective if C has a section over S .

2.3. If T is an S-scheme $S' \to S$ an fffp-extension, $C' = C \times_S S'$, $T' = T \times_S S'$, then $\mathrm{Pic}^n_{C/S}(T) \to \mathrm{Pic}^n_{C'/S'}(T')$ is injective. Combining these facts with 1.2. we get

2.4. Let U be a schematically dense open subscheme of the S-scheme T , then $\mathrm{Pic}^n_{C/S}(T) \to \mathrm{Pic}^n_{C/S}(U)$ is injective. The objective of the rest of this chapter is to prove

<u>Theorem 2.5</u>. $\mathrm{Pic}^n_{C/S}$ is representable locally in the fffp-topology.

<u>Remark 2.6</u>. It suffices to prove 2.5. for $n = 0$, namely, if there exists on C an invertible sheaf of rank n then Pic^0 and Pic^n are isomorphic, and after an fffp-extension of S there does exist such a sheaf as it follows from III.1.1. which in fact remains valid for any integer $g \geq 1$, and not only for g = genus, as if follows from the proof of III.1.1.

IV.3. Construction of a group scheme.

In this section we fix an invertible sheaf D of degree g on C. Such a D exists after an fffp-extension S, according to III.1.1.

Notation: If $f : T_1 \to T_2$ is an S-morphism and $E \in Pic^n_{C/S}(T_2)$ then f^*E is short for $Pic^n_{C/S}(f)(E)$. The same notation will be used if f is an S-dominant S-rational map and $E \in Pic^n(U_2)$ where U_2 is an S-dense open of T_2, f^*E is then an element of $Pic^n(U_1)$ where U_1 is an unspecified S-dense open of T_1. This abuse of notation is justified by the seesaw principle 2.4.

According to III.4.10 there exists a unique S-dominant S-rational map

3.1. $$a : C^{(g)} \times_S C^{(g)} \to C^{(g)}$$

such that with the notation of III.4.10

3.2. $$a^*\mathcal{L} = p_1^*\mathcal{L} \otimes p_2^* \mathcal{L} \otimes \pi^*D^{-1}$$

where $\pi : C^{(g)} \to S$ is the structure map.

It is convenient to introduce

3.3. $$\mathcal{L}_0 = \mathcal{L} \otimes \pi^* D^{-1}$$

$\mathcal{L}_0$ is now an element of $Pic^0_{C/S}(U)$ for some S-dense open of $C^{(g)}$. 3.3. can now be rewritten

3.4. $$a^*\mathcal{L}_0 = p_1{}^*\mathcal{L}_0 \otimes p_2{}^*\mathcal{L}_0$$

It follows from III.4.10 that a is commutative and associative in the obvious sense.

Lemma 3.5. Let $\Gamma_a : C^{(g)} \times C^{(g)} \to C^{(g)} \times C^{(g)} \times C^{(g)}$ denote the S-rational map which is the graph of $a : C^{(g)} \times C^{(g)} \to C^{(g)}$. Then $p_{13} \circ \Gamma_a$ and $p_{23} \circ \Gamma_a$ are S-dominant rational maps which are isomorphisms in the category of S-dominant S-rational maps.

Proof: Let us treat $p_{13} \circ \Gamma_a$.

$$p_{13} \circ \Gamma_a = (p_1, a) : C^{(g)} \times C^{(g)} \to C^{(g)} \times C^{(g)}$$

which is S-dominant by III.4.11. By III.4.10 we have a unique S-dominant S-rational map $s : C^{(g)} \times C^{(g)} \to C^{(g)}$ such that $s^*\mathcal{L}_0 = p_2{}^*\mathcal{L}_0 \otimes p_1{}^*\mathcal{L}_0$. The map $(p_1, s) : C^{(g)} \times C^{(g)} \to C^{(g)} \times C^{(g)}$ is again S-dominant by III.4.11. It is now straight forward to check by means III.4.10 that (p_1, s) and (p_1, a) are inverses to each other in the category of S-dominant S-rational maps.

Q.E.D.

IV.4. Proof of the theorem on representability.

In this section we assume given the following data over S

$$(D,J,U,i,j)$$

where D is an invertible sheaf on C of degree g, hence we have $a : C^{(g)} \times C^{(g)} \to C^{(g)}$ and $\mathcal{L}_0$ as in IV.3. J is a smooth, commutative group scheme whose geometric fibers are connected, $i : U \to J$ an open S-dense inversion, such that the two rational maps from $U \times U$ into U induced by the addition on J and by $a : C^{(g)} \times C^{(g)} \to C^{(g)}$ coincide.

Such data exists according to IV.3 and [S.G.A.D.] 5B exp. XVIII Th. 3.7.

The aim of this section is to prove

4.1. There exist $\mathcal{L} \in \mathrm{Pic}^0(J)$ such that $i^*\mathcal{L} = \mathcal{L}_0$. Such an $\mathcal{L}$ is unique by the seesaw principle.

4.2. The pair $(J,\mathcal{L})$ represents $\mathrm{Pic}^0_{C/S}$.

Proof of 4.1: Consider the map $\alpha : U \times U \to J$ which is the composite of $i \times i$ and the addition on J. This map is fffp namely, the addition map is that [S.G.A.D.] 2A exp. VI, to prove that α is surjective it suffices to treat the case where S is the spectrum of an alg. closed field h ; let $s \in J(h)$ then $(s - U) \cap U \neq \emptyset$. By 3.4 we want $\mathcal{L} \in \mathrm{Pic}(J)$ such that $\alpha^*\mathcal{L} = p_1{}^*\mathcal{L}_0 \otimes p_2{}^*\mathcal{L}_0$. Since Pic^0 is an fffp-sheaf it suffices to prove

$$\alpha_1{}^*(p_1{}^*\mathcal{L}_0 \otimes p_2{}^*\mathcal{L}_0) = \alpha_2{}^*(p_1{}^*\mathcal{L}_0 \otimes p_2{}^*\mathcal{L}_0)$$

where α_1 and α_2 denotes the two projections from $(U\times U)\times_J(U\times U)$ onto $U\times U$. Consider the following catesian diagrams

$$\begin{array}{ccc} U\times U\times U\times U & \to & J\times J\times J\times J \\ \uparrow & & \uparrow \\ (U\times U)\times_J(U\times U) & \to & (J\times J)\times_J(J\times J) \end{array}$$

and note that $(J\times J)\times_J(J\times J)$ can be identified with the the graph of the map $J\times J\times J \to J$ which sends (a,b,c) into $(a,b,-c+a+b)$. It follows that $(U\times U)\times_J(U\times U)$ can be identified with the graph of the S-rational map $U\times U\times U \to U$ induced by $(a,b,c) \to -c+a+b$.
By the seesaw principle it suffices to prove that

$$\beta_1{}^*(p_1{}^*\mathcal{L} \otimes p_2{}^*\mathcal{L}) = \beta_2{}^*(p_1{}^*\mathcal{L} \otimes p_2{}^*\mathcal{L})$$

where $\beta_1 : U\times U\times U \to U\times U$ is the map $(a,b,c) \to (a,b)$ and $\beta_2 : U\times U\times U \to U\times U$ the S-rational map induced by $(a,b,c) \to (a,b,-c+a+b)$. This is easy to check once the following facts are recorded: Let s (resp. a) denote the S-rational map $U\times U \to U$ induced by $(x,y) \to x-y$ (resp. $(x,g) \to (x+y)$ then

4.3.
$$a^*\mathcal{L}_0 = p_1{}^*\mathcal{L}_0 \otimes p_2{}^*\mathcal{L}_0$$
$$s^*\mathcal{L}_0 = p_1{}^*\mathcal{L}_0 \otimes p_2{}^*\mathcal{L}_0{}^{-1}$$

The first relation is nothing but 3.4. Let $r : U\times U \to U\times U$ denote the S-rational map induced by $(x,y) \to (x+y,y)$. By seesaw principle it suffices to prove that $r^*s^*\mathcal{L}_0 = r^*(p_1{}^*\mathcal{L}_0 \otimes p_2{}^*\mathcal{L}_0{}^{-1})$, which is clear.

Q.E.D.

<u>Remark 4.4</u>. The reader is invited to formulate an abstract lemma to which the previous proof applies.

<u>Remark 4.5</u>. The element $\mathcal{L} \in \mathrm{Pic}^0(J)$ satisfies the following relations

$$a^*\mathcal{L} = p_1{}^*\mathcal{L} \otimes p_2{}^*\mathcal{L} \qquad (a : (x,y) \to x+y)$$
$$s^*\mathcal{L} = p_1{}^*\mathcal{L} \otimes p_2{}^*\mathcal{L}^{-1} \qquad (s : (x,y) \to x-y)$$

as it follows from 4.3 and the seesaw principle.

<u>Proof of 4.2</u>: $\mathcal{L} \in \mathrm{Pic}^0(J)$ defines a map $\mathrm{Hom}_S(\ ,J) \to \mathrm{Pic}^0$ namely, to $f : I \to J$ we associate $f^*\mathcal{L} \in \mathrm{Pic}^0(T)$

<u>1^0 $\mathrm{Hom}_S(\ ,J) \to \mathrm{Pic}^0$ is injective</u>:

Since everything we have dealt with is preserved

under base-change it suffices to prove that

$$J(S) = \mathrm{Hom}_S(S,T) \to \mathrm{Pic}^0(S)$$

is injective. For this we may assume that C has an S-section as it follows from 2.3 and now we can identify $\mathrm{Pic}^0_{C/S}$ with "Pic naif". We may shrink U a little as to get that $U(T) \to \mathrm{Pic}^0(T)$ is injective for all S-schemes T, namely just replace U by $W = \{w \in D^{(g)} \cap U \mid h^1(\mathcal{L}_w) = 0\}$ where $D^{(g)}$ is the g-fold symmetric product of the smooth locus of $C \to S$. W is open by the upper semicontinuity theorem and S-dense by III.3.1. That W will work follows from the proof of III.4.3. According to 4.5 it suffices now to prove that for $r,s \in J(S)$ there exist an fffp-extension $S' \to S$ and $t \in J(S')$ such that $r+t$ and $s+t \in U(S')$. For this it suffices to prove that the map $J \times U \times U \to J \times J$ which sends (j,u,v) into $(j+u, j+v)$ is fffp, since we then locally in the fffp-topology can lift the section (r,s). That $(j,u,v) \to (j+u,j+v)$ is flat follows from the fact that the addition on J is flat, that it is surjective is easy to establish in the case where $S = \mathrm{Spec}(\bar{k})$, which will suffice.

Q.E.D.

2^0 <u>$\mathrm{Hom}_S(\ ,J) \to \mathrm{Pic}^0$ is surjective</u>:

By functoriality we only have to show that $J(S) \to \mathrm{Pic}^0(S)$ is surjective. Given $L \in \mathrm{Pic}^0(S)$, if we can find an fffp-extension $f : S' \to S$ and $t \in J(S')$

such that $t^*\mathcal{L} = L' = f^*L$ then we may descent t by the fact that $J(\ ) \to \mathrm{Pic}^0(\ )$ is injective and the theorem on fffp-descent of morphisms [S.G.A.] 60- 61 exp. VIII Th. 5.2.

We may now assume that C has a section over S so that we can identify Pic^0 with "Pic naif". According to theorem III.4.5 we can find S-dominant S-rational map $r_L : C^{(g)} \to C^{(g)}$ such that

$$r_L^*(\mathcal{L}_0) = \mathcal{L}_0 \otimes \pi^* L$$

where $\pi : C^{(g)} \to S$ denotes the structure map. Choose an S-dense open U_L of $C^{(g)}$ such that $U_L \subseteq U$ and $r_L(U_L) \subseteq U$. Extending the base we may assume that U_L has a section, s say. Let $i : U_L \to J$ denote the inclusion,

$$t = i \circ r_L \circ s - i \circ s \in J(S)$$

satisfy the property $t^*\mathcal{L} = L$ as we will verify by means of 4.5:

$$\begin{aligned} t^*\mathcal{L} &= s^* r_L^* i^* \mathcal{L} \otimes s^* i^* \mathcal{L} \\ &= s^* r_L^* \mathcal{L}_0 \otimes s^* \mathcal{L}_0^{-1} \\ &= s^* \mathcal{L}_0 \otimes s^* \pi^* L \otimes s^* \mathcal{L}_0^{-1} = L \end{aligned}$$

Q.E.D.

Appendix. The closed immersion of a family curves into its Picard scheme.

We offer a proof of the following

Proposition. Let S be a noetherian scheme, $f : C \to S$ a flat and proper morphism whose geometric fibers are smooth connected curves of genus $g \geq 1$ and let $J^1_{C/S}$ represent $\mathrm{Pic}^1_{C/S}$. Then the natural map $C \to J^1_{C/S}$ is a closed immersion.

The proof is based on the following

Lemma. Let S be a noetherian scheme $g : X \to Y$ an S-morphism where X is proper over S and Y is separated and of finite type over S . If g is a monomorphism in the category of schemes of finite type over S then g is a closed immersion.

Proof: g is proper by [E.G.A.] II 5.4.3 so g is finite by Chevalley's theorem [E.G.A.] III 4.4.2. It now suffices to prove : let $T \to S$ be finite morphism which is a monomorphism in the category of finite type S-schemes then $T \to S$ is a closed immersion. For this we may assume S is affine, $S = \mathrm{Spec}\ A$ say, and that $T = \mathrm{Spec}\ B$. It follows that $B \otimes_A B$ is A-isomorphic to B and hence by

Nakayamas lemma that $A \to B$ is surjective.

Q.E.D.

Proof of the proposition:

It suffices to prove that for any finite type S-scheme T

$$C(T) \to \mathrm{Pic}^1_{C/S}(T)$$

is injective. By naturality and base-change it suffices to prove this for $T = S$ and by IV.2.2 and IV.2.3 we can replace $\mathrm{Pic}^1_{C/S}$ by "Pic naif".

Now let $s_i \in C(S)$ $i = 1,2$ and let L_i denote the invertible sheaf given by

$$0 \to L_i^{-1} \to \mathcal{O}_C \to s_{i*}\mathcal{O}_S \to 0$$

and suppose that $L_1 = L_2 \otimes f^*L$, where L is an invertible sheaf on S. We first prove that f_*L_i is locally free and that for all $s \in S$ the restriction map

$$r_s : f_*L_i \otimes k(s) \to H^0(X_s, L_{is})$$

is an isomorphism. Note first that $H^0(X_s, L_{is})$ is one dimensional since the fibers are non rational curves and that the global section t_i of L_i corresponding to the inclusion $L_i^{-1} \to \mathcal{O}_C$ has non zero image by r_s. By IV.1.4 we can conclude that f_*L_i is locally free and that r_s is an isomorphism. It follows from Nakayamas lemma that

$t_i : \mathcal{O}_S \to f_* L_i$ is an isomorphism. Since $L_1 = L_2 \otimes f^* L$ it follows that $L = f_* f^* L$ is trivial and hence $L_1 = L_2$. Consider t_1 and t_2 as global sections of $f_* L_1$. As such they will only differ by a global S-unit as it follows from the fact that $f_* L_1 \cong \mathcal{O}_S$ and that t_1 and t_2 has a non trivial restriction to all fibers.

Q.E.D.

BIBLIOGRAPHY

G] M. Auslander and O. Goldman, "The Brauer group of a commutative ring", Trans. Amer. Math. Soc., 1960 (97), pp. 367-409.

A. Grothendieck, "Le groupe de Brauer, I", Séminaire Bourbaki Mai 1965, n° 290.

G.A.] A. Grothendieck, "Élements de géometrie algébrique", Publ. Math. de l'Inst. des Hautes Ét. Sci., Paris, no. 4,8,11,17,20,24,28,32 (1960 ff).

G.A.] A. Grothendieck, "Séminaire de Géométrie Algébrique", notes polycopiés, I.H.E.S., Paris, 1960-61.

D.T.E.] A. Grothendieck, "Les Schemes de Picard. Théorème d'existence". Seminaire Bourbaki, t. 14, 1961/62, n° 232.

G.A.D.] "Seminaire de Géométrie Algèbrique", notes polycopiés, I.H.E.S., Paris 1963-64.

D. Lazard, "Autour de la platitude", Bull. Soc. Math. France, 1969 (97) pp. 81-128.

D. Mumford, "Introduction to abelian varieties", Oxford University Press, 1970.

] D. Mumford, "Lectures on curves on an algebraic surface", Annals of Mathematics Studies no. 59, Princeton University Press 1966.

] M. Raynaud, "Faisceaux amples sur les schémas et les espace homogènes", Lecture notes in mathematics, no. 119, Springer-Verlag, Berlin, 1970.

] M. Rosenlicht, "Equivalence relations on algebraic curves", Ann. of Maths., 1952 (56) pp. 169-191.

] M. Rosenlicht, "Generalized Jacobian varieties", Ann. of Maths., 1957 (66) pp. 505-530.

J.-P. Serre, "Groupes Algébriques et Corps de Classes", Hermann, Paris, 1959.

A. Weil, "Variétés abéliennes et courbes algébriques", Hermann, Paris, 1948.

Lecture Notes in Mathematics

Bisher erschienen/Already published

Vol. 1: J. Wermer, Seminar über Funktionen-Algebren. IV, 30 Seiten. 1964. DM 3,80 / $ 1.10

Vol. 2: A. Borel, Cohomologie des espaces localement compacts d'après. J. Leray. IV, 93 pages. 1964. DM 9,– / $ 2.60

Vol. 3: J. F. Adams, Stable Homotopy Theory. Third edition. IV, 78 pages. 1969. DM 8,– / $ 2.20

Vol. 4: M. Arkowitz and C. R. Curjel, Groups of Homotopy Classes. 2nd. revised edition. IV, 36 pages. 1967. DM 4,80 / $ 1.40

Vol. 5: J.-P. Serre, Cohomologie Galoisienne. Troisième édition. VIII, 214 pages. 1965. DM 18,– / $ 5.00

Vol. 6: H. Hermes, Term Logic with Choise Operator. III, 55 pages. 1970. DM 6,– / $ 1.70

Vol. 7: Ph. Tondeur, Introduction to Lie Groups and Transformation Groups. Second edition. VIII, 176 pages. 1969. DM 14,– / $ 3.80

Vol. 8: G. Fichera, Linear Elliptic Differential Systems and Eigenvalue Problems. IV, 176 pages. 1965. DM 13,50 / $ 3.80

Vol. 9: P. L. Ivănescu, Pseudo-Boolean Programming and Applications. IV, 50 pages. 1965. DM 4,80 / $ 1.40

Vol. 10: H. Lüneburg, Die Suzukigruppen und ihre Geometrien. VI, 111 Seiten. 1965. DM 8,– / $ 2.20

Vol. 11: J.-P. Serre, Algèbre Locale. Multiplicités. Rédigé par P. Gabriel. Seconde édition. VIII, 192 pages. 1965. DM 12,– / $ 3.30

Vol. 12: A. Dold, Halbexakte Homotopiefunktoren. II, 157 Seiten. 1966. DM 12,– / $ 3.30

Vol. 13: E. Thomas, Seminar on Fiber Spaces. IV, 45 pages. 1966. DM 4,80 / $ 1.40

Vol. 14: H. Werner, Vorlesung über Approximationstheorie. IV, 184 Seiten und 12 Seiten Anhang. 1966. DM 14,– / $ 3.90

Vol. 15: F. Oort, Commutative Group Schemes. VI, 133 pages. 1966. DM 9,80 / $ 2.70

Vol. 16: J. Pfanzagl and W. Pierlo, Compact Systems of Sets. IV, 48 pages. 1966. DM 5,80 / $ 1.60

Vol. 17: C. Müller, Spherical Harmonics. IV, 46 pages. 1966. DM 5,– / $ 1.40

Vol. 18: H.-B. Brinkmann und D. Puppe, Kategorien und Funktoren. XII, 107 Seiten, 1966. DM 8,– / $ 2.20

Vol. 19: G. Stolzenberg, Volumes, Limits and Extensions of Analytic Varieties. IV, 45 pages. 1966. DM 5,40 / $ 1.50

Vol. 20: R. Hartshorne, Residues and Duality. VIII, 423 pages. 1966. DM 20,– / $ 5.50

Vol. 21: Seminar on Complex Multiplication. By A. Borel, S. Chowla, C. S. Herz, K. Iwasawa, J.-P. Serre. IV, 102 pages. 1966. DM 8,– /$ 2.20

Vol. 22: H. Bauer, Harmonische Räume und ihre Potentialtheorie. IV, 175 Seiten. 1966. DM 14,– / $ 3.90

Vol. 23: P. L. Ivănescu and S. Rudeanu, Pseudo-Boolean Methods for Bivalent Programming. 120 pages. 1966. DM 10,– / $ 2.80

Vol. 24: J. Lambek, Completions of Categories. IV, 69 pages. 1966. DM 6,80 / $ 1.90

Vol. 25: R. Narasimhan, Introduction to the Theory of Analytic Spaces. IV, 143 pages. 1966. DM 10,– / $ 2.80

Vol. 26: P.-A. Meyer, Processus de Markov. IV, 190 pages. 1967. DM 15,– / $ 4.20

Vol. 27: H. P. Künzi und S. T. Tan, Lineare Optimierung großer Systeme. VI, 121 Seiten. 1966. DM 12,– / $ 3.30

Vol. 28: P. E. Conner and E. E. Floyd, The Relation of Cobordism to K-Theories. VIII, 112 pages. 1966. DM 9,80 / $ 2.70

Vol. 29: K. Chandrasekharan, Einführung in die Analytische Zahlentheorie. VI, 199 Seiten. 1966. DM 16,80 / $ 4.70

Vol. 30: A. Frölicher and W. Bucher, Calculus in Vector Spaces without Norm. X, 146 pages. 1966. DM 12,– / $ 3.30

Vol. 31: Symposium on Probability Methods in Analysis. Chairman. D. A. Kappos.IV, 329 pages. 1967. DM 20,– / $ 5.50

Vol. 32: M. André, Méthode Simpliciale en Algèbre Homologique et Algèbre Commutative. IV, 122 pages. 1967. DM 12,– / $ 3.30

Vol. 33: G. I. Targonski, Seminar on Functional Operators and Equations. IV, 110 pages. 1967. DM 10,– / $ 2.80

Vol. 34: G. E. Bredon, Equivariant Cohomology Theories. VI, 64 pages. 1967. DM 6,80 / $ 1.90

Vol. 35: N. P. Bhatia and G. P. Szegö, Dynamical Systems. Stability Theory and Applications. VI, 416 pages. 1967. DM 24,– / $ 6.60

Vol. 36: A. Borel, Topics in the Homology Theory of Fibre Bundles. VI, 95 pages. 1967. DM 9,– / $ 2.50

Vol. 37: R. B. Jensen, Modelle der Mengenlehre. X, 176 Seiten. 1967. DM 14,– / $ 3.90

Vol. 38: R. Berger, R. Kiehl, E. Kunz und H.-J. Nastold, Differentialrechnung in der analytischen Geometrie IV, 134 Seiten. 1967 DM 12,– / $ 3.30

Vol. 39: Séminaire de Probabilités I. II, 189 pages. 1967. DM 14,– / $ 3.90

Vol. 40: J. Tits, Tabellen zu den einfachen Lie Gruppen und ihren Darstellungen. VI, 53 Seiten. 1967. DM 6.80 / $ 1.90

Vol. 41: A. Grothendieck, Local Cohomology. VI, 106 pages. 1967. DM 10,– / $ 2.80

Vol. 42: J. F. Berglund and K. H. Hofmann, Compact Semitopological Semigroups and Weakly Almost Periodic Functions. VI, 160 pages. 1967. DM 12,– / $ 3.30

Vol. 43: D. G. Quillen, Homotopical Algebra. VI, 157 pages. 1967. DM 14,– / $ 3.90

Vol. 44: K. Urbanik, Lectures on Prediction Theory. IV, 50 pages. 1967. DM 5,80 / $ 1.60

Vol. 45: A. Wilansky, Topics in Functional Analysis. VI, 102 pages. 1967. DM 9,60 / $ 2.70

Vol. 46: P. E. Conner, Seminar on Periodic Maps. IV, 116 pages. 1967. DM 10,60 / $ 3.00

Vol. 47: Reports of the Midwest Category Seminar I. IV, 181 pages. 1967. DM 14,80 / $ 4.10

Vol. 48: G. de Rham, S. Maumary et M. A. Kervaire, Torsion et Type Simple d'Homotopie. IV, 101 pages. 1967. DM 9,60 / $ 2.70

Vol. 49: C. Faith, Lectures on Injective Modules and Quotient Rings. XVI, 140 pages. 1967. DM 12,80 / $ 3.60

Vol. 50: L. Zalcman, Analytic Capacity and Rational Approximation. VI, 155 pages. 1968. DM 13.20 / $ 3.70

Vol. 51: Séminaire de Probabilités II. IV, 199 pages. 1968. DM 14,– / $ 3.90

Vol. 52: D. J. Simms, Lie Groups and Quantum Mechanics. IV, 90 pages. 1968. DM 8,– / $ 2.20

Vol. 53: J. Cerf, Sur les difféomorphismes de la sphère de dimension trois ($\Gamma_4 = O$). XII, 133 pages. 1968. DM 12,– / $ 3.30

Vol. 54: G. Shimura, Automorphic Functions and Number Theory. VI, 69 pages. 1968. DM 8,– / $ 2.20

Vol. 55: D. Gromoll, W. Klingenberg und W. Meyer, Riemannsche Geometrie im Großen. VI, 287 Seiten. 1968. DM 20,– / $ 5.50

Vol. 56: K. Floret und J. Wloka, Einführung in die Theorie der lokalkonvexen Räume. VIII, 194 Seiten. 1968. DM 16,– / $ 4.40

Vol. 57: F. Hirzebruch und K. H. Mayer, O (n)-Mannigfaltigkeiten, exotische Sphären und Singularitäten. IV, 132 Seiten. 1968. DM 10,80/ $ 3.00

Vol. 58: Kuramochi Boundaries of Riemann Surfaces. IV, 102 pages. 1968. DM 9,60 / $ 2.70

Vol. 59: K. Jänich, Differenzierbare G-Mannigfaltigkeiten. VI, 89 Seiten. 1968. DM 8,– / $ 2.20

Vol. 60: Seminar on Differential Equations and Dynamical Systems. Edited by G. S. Jones. VI, 106 pages. 1968. DM 9,60 / $ 2.70

Vol. 61: Reports of the Midwest Category Seminar II. IV, 91 pages. 1968. DM 9,60 / $ 2.70

Vol. 62: Harish-Chandra, Automorphic Forms on Semisimple Lie Groups X, 138 pages. 1968. DM 14,– / $ 3.90

Vol. 63: F. Albrecht, Topics in Control Theory. IV, 65 pages. 1968. DM 6,80 / $ 1.90

Vol. 64: H. Berens, Interpolationsmethoden zur Behandlung von Approximationsprozessen auf Banachräumen. VI, 90 Seiten. 1968. DM 8,– / $ 2.20

Vol. 65: D. Kölzow, Differentiation von Maßen. XII, 102 Seiten. 1968. DM 8,– / $ 2.20

Vol. 66: D. Ferus, Totale Absolutkrümmung in Differentialgeometrie und -topologie. VI, 85 Seiten. 1968. DM 8,– / $ 2.20

Vol. 67: F. Kamber and P. Tondeur, Flat Manifolds. IV, 53 pages. 1968. DM 5,80 / $ 1.60

Vol. 68: N. Boboc et P. Mustață, Espaces harmoniques associés aux opérateurs différentiels linéaires du second ordre de type elliptique. VI, 95 pages. 1968. DM 8,60 / $ 2.40

Vol. 69: Seminar über Potentialtheorie. Herausgegeben von H. Bauer. VI, 180 Seiten. 1968. DM 14,80 / $ 4.10

Vol. 70: Proceedings of the Summer School in Logic. Edited by M. H. Löb. IV, 331 pages. 1968. DM 20,– / $ 5.50

Vol. 71: Séminaire Pierre Lelong (Analyse), Année 1967 – 1968. VI, 190 pages. 1968. DM 14,– / $ 3.90

Bitte wenden / Continued

274